A Study On Fixed Point Theory With Applications

Edited and Compiled by

Ph. Suranjoy Singh, Gold Medalist,
M.Sc. Mathematics, University of Delhi.

2024 Culturea
xte et illustration de couverture : © domaine public (open access licence CC0 1.0 universel)
tion et mise en page : Culturea (Hérault, 34)
rouvez notre catalogue sur http://culturea.fr/
ntact : infos@culturea.fr
primé en Allemagne par Books on Demand Gmbh
sign typographique et layout : Derek Murphy et Reedsy
3N : 9791041985661
te de parution : février 2024
livre :
a été composé avec une police open source libre de droits dessinée sur Open Fondry https://open-foundry.com/
est édité en libre access sous licence CC0 (Creative Commons Zero)
fin de conserver l'oeuvre au plus près des caractéristiques du domaine public.
offre la possibilité à son lecteur de partager, dupliquer ou distribuer son contenu, sans restriction ni réserve.
permet de replanter un arbre. SHS Éditions soutient Plantons Pour l'Avenir, fonds de dotation pour le reboisement des forêts françaises
0 projets de reboisement de forêts soutenus depuis 2014 à travers la France https://www.plantonspourlavenir.fr/

Preface

Fixed Point Theory is an attractive and interesting subject with a large number of applications in various fields of mathematics and other branches of science

The main intention of writing this book is, as the topic of the book "A Study On Fixed Point Theory With Applications" implies, is to give a rough idea of the basic types, some important theorems and a few common applications of Fixed Point Theorey and also, to enhance my career as a mathematician in the field of fixed point theory. The book will serve good for the beginners in the field of fixed point in the similar manner as I was beniffited from the valuable contents provided in the book.

In writing this book, the works of standard authors have been a great help and I am greatly indebted to them. Above all, research papers and articles of eminent researchers and authors and internet too were very useful while completing this book. Last but not the least, the support my wife Mrs. H. Dayapati, M.Sc. Math, has been the pillar of my strength in my work all through.

I express my words of thanks to Freeditorial Publisher and Library for taking the tough of distributing this book.

Any errors that may appear are, of course, my responsibility. I warmly welcome any suggestion or correction from the valuable readers of this book.

Ph. Suranjoy Singh

Contents

1. Introduction 1

2. Topological Fixed Point Theory 3
- 2.1 Introduction and Background 3
- 2.2 The Brower's Fixed point Theorem 9
- 2.3 Sperner's Lemma and Brower's Fixed Point Theorem Theorem 31
- 2.4 Generalization 38
- 2.5 Multifunctions and Kakutani's Theorem 52
- 2.6 Applications 57

3. Metric Fixed Point Theory 75
- 3.1 Introduction 75
- 3.2 Banach's Contraction Principle, extensions and converse 75
- 3.3 Metric Fixed Point Theory in Banach Spaces 88
- 3.4 Metric Fixed Point Theory in Metric Spaces 95
- 3.5 Applications 102

4. Order-Theoretic Fixed Point Theory 121
- 4.1 Introduction 121
- 4.2 Completeness Condition for Posets 121
- 4.3 Iterative Fixed Point Theorems 126
- 4.4 The Tarski's Fixed Point Theorem 133
- 4.5 The Abian-Brown Fixed Point Theorem 135

Thanks for reading 138

Bibliography 139

Chapter 1
Introduction

Fixed point : Let X be a set and let $T: X \to X$ be a function that maps X into itself. (Such a function is often called an operator, a transformation or a transform on X and the notation Tx is often used in place of $T(x)$. A fixed point of T is an element $x \in X$ for which $T(x) = x$.

Example: Let X be the two-element set $\{x, y\}$. The function $f: X \to X$ defined by $f(x) = y$ and $f(y) = x$ has no fixed point but the other three functions that map X into itself each have one or two fixed points.

N.B. : The definition of a fixed point requires no structure on either the set X or the function T.

***Common fixed point*:** Let X be a nonempty set and $\mathcal{F}$ a nonempty collection of self-maps on X. By a fixed point of $\mathcal{F}$ or by a common fixed point of the maps in $\mathcal{F}$, we mean an element x of X such that $x = f(x)$ for each $f \in \mathcal{F}$. The set of all fixed points of $\mathcal{F}$ is denoted as Fix($\mathcal{F}$):

Fixed point theorems : The theorems concerning the existence of fixed points and their properties are called fixed point theorems. Informally speaking, fixed point theory is a branch of mathematics that attempts to identify all self-maps (or self-correspondences) under which at least one element is left invariant.

Fixed point theory is divided into three major types :

i) Topological Fixed Point Theory
ii) Metric Fixed Point Theory
iii) Order-Theoretic Fixed Point Theory

Historically, the boundary lines among the three types were respectively defined by the discoveries of following three major theorems :

i) Brower's Fixed Point Theorem (1910)
ii) Banach's Fixed Point Theorem (1922)
iii) Tarski's Fixed Point Theorem (1955)

Fixed point theory is an attractive and interesting subject with a large number of applications in various fields of mathematics and other branches of science. Fixed point theory has become not only a field with a huge development but also a very helpful means for solving various problems in different fields of mathematics. Fixed point theorems are used for proving the existence and uniqueness to various mathematical models like differential, integral and partial differential equations and variational inequalities etc., representing phenomena arising in different fields such as steady state temperature distribution, chemical equations, neutron transport theory, economic theories, epidemics and flow of fluids. Above all, it as also useful in the field of computer science, image processing, artificial intelligence, decision making, population dynamics, computer science, operational research, industrial engineering, pattern recognition, medicine, group health underwriting, management and many others. Only a few common applacations are provided in this book.

Chapter 2
Topological Fixed Point Theory

2.1 Introduction and background

***Definition* :** *Let X be a nonempty set and τ a collection of subsets of X. Then, τ is said to be a topology on X if the following conditions are satisfied:*
(i) $\phi \in \tau$ and $X \in \tau$,
(ii) τ is closed under arbitrary unions,
(iii) τ is closed under finite intersections.

The ordered pair (X, τ) is called topological space. The members of τ are called τ-open sets or simply open sets.

Topological fixed point theory plays an important role in the development of algebraic topology during the early decades of the twentieth century. Bolzano, Brouwer and Borsuk are the early pioneers in the field of topological fixed point theory. The first topological fixed point theorem is the *Intermediate Value Theorem*, stated as : *Let $f : [-1, 1] \rightarrow \boldsymbol{R}$ be a continuous function such that $f(-1) < 0$ and $f(1) > 0$. Then, there exists a point $c \in [-1, 1]$ such that $f(c) = 0$* which was first proved by Bernard Bolzano (1781 - 1848) in 1817.

2.1.1 The Topological Fixed Point Property

A fixed point of a function $f : X \rightarrow X$ is an element $x \in X$ that satisfies $f(x) = x$. Given a set X, it is possible to ask what types of functions on X have a fixed point. Alternatively, we could consider a class of functions and investigate the kinds of sets on which a function in our class will have a fixed point. It is the latter course of investigation that is the main subject here. A set X is said to have the **topological fixed point property** (*tfpp*) provided every continuous self– map on X has a fixed point. Clearly, the space $\boldsymbol{R}^n$ does not have the topological fixed point property. It is not hard to see the closed unit interval does have the fixed point property.

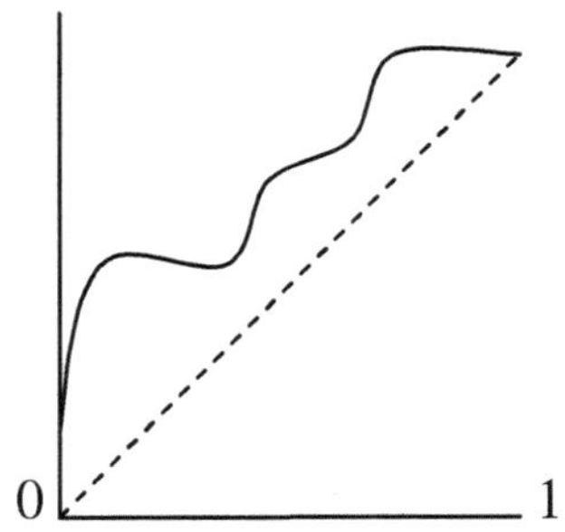

Figure : The fixed point property on [0, 1].

As the above picture illustrates, if $f : [0, 1] \rightarrow [0, 1]$ is a continuous function with no fixed point in $[0,1)$, then $f(0) \neq 0$. As we let x vary continuously from 0 to 1, by our assumption and continuity of f, $f(x)$ must always stay between x and 1. This forces $f(1) = 1$.

In general, it is not known exactly what type of sets possess the *tfpp*. At the very least, we might expect that such a set would have to contain its limit points. Otherwise, a continuous function might be able to shift each element of X closer to a missing limit point. Thus, it seems reasonable that compactness be a required property. However, this is clearly not sufficient. For example, the set $[0, 1] \cup [2, 3]$ does not have the *tfpp*. In light of this, we might also ask that our set have no "holes" in it. For, in this case, we might consider a rotation around the missing set. Thus, it might be prudent to restrict ourselves to contractible sets. That is, sets that can be continuously deformed to a point. Of course, this property alone would not guarantee a fixed point for f. For example, (0, 1) does not have the *tfpp*. The next logical step would be to consider whether sets with both properties, sets that are compact and contractible, have the *tfpp*. This question was posed by Borsuk in 1932, and remained open for more than 20 years. It was answered in the negative when Kinoshita gave a beautiful example of a compact, contractible set without the topological fixed point property. What we need is something a little stronger than contractibility. Any convex set is contractible. It turns out that compact and convex is sufficient to ensure the existence of fixed points. Also, among convex sets, compactness is necessary.

However, there do exist sets that are non convex that do have the *tfpp.* One interesting example is the sin(1/*x*) circle. This set consists of the closure of the set $\{(x, y) : y = \sin(1/x\ ),\ 0 < x \leq 1/\pi\}$ together with an arc joining the points (0, 1) and ($1/\pi$, 0).

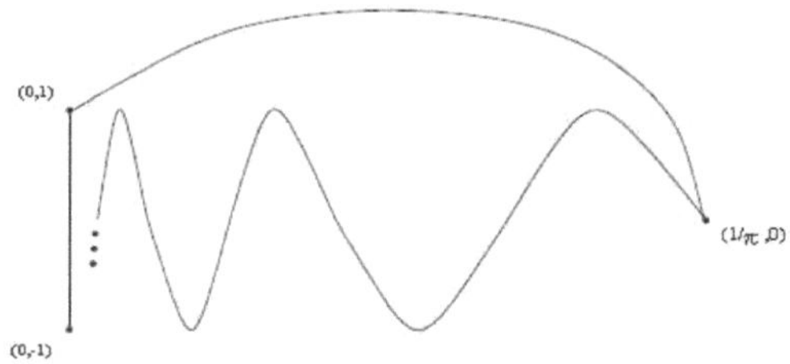

Figure : The sin(1/*x)* circle.

An argument similar to that above for the unit interval shows that this set has the *tfpp*. This "nowhere left to go" reasoning is sometimes referred to as the "dog chasing a rabbit" argument.

The following is an interesting example of a set in $\boldsymbol{R}^2$ that is neither convex nor compact but still has the topological fixed point proper-ty.

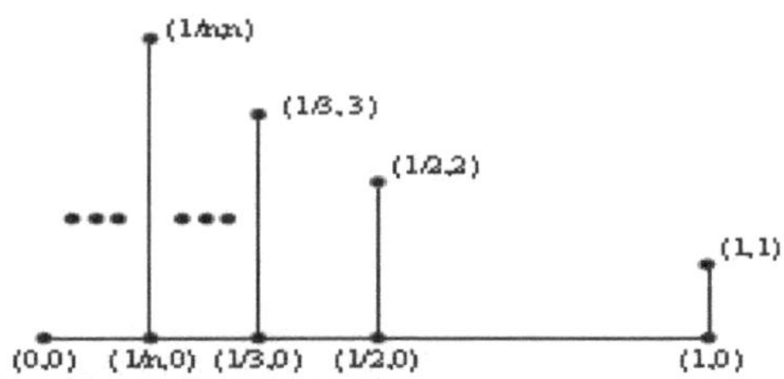

Figure : An unbounded, non closed set with the *tfpp.*

The figure consists of a base segment $X_0 = [0,\ 1]$ on the *x*–axis with vertical segments X_n starting at the point $(1/n,\ 0)$ and extending to $(1/n,\ n)$. This set is neither closed nor bounded, thus certainly not compact. To see that X has the fixed point property, let $f : X \rightarrow X$ be continuous. Denote by x_n the point $(1/n, 0)$. There are three cases to consider. First, if there is an $n \in \boldsymbol{N}$ with $f(x_n) = x_n$, then we are done. Else, suppose there is an n such that $f(x_n) \in X_n \setminus \{x_n\}$. Now, let r be a retraction of X onto X_n. Then, $r \circ f : X_n \rightarrow X_n$ is continuous and thus, has a fixed point, say x_o. By our assumption, x_o cannot be x_n and hence $x_o = r(f(x_o)) = f(x_o)$. The final case to consider is that for every $n \in \boldsymbol{N}$, $f(x_n)$ is not in X_n. Then, let r be the retraction of X onto X_0 and a similar argument as in the previous case gives us a fixed point for f. Thus, among non convex sets, compactness and contractibility do not have a direct relationship with the topological fixed point property. In general, it is not known what types on non convex sets have the property.

Remark 1 **:** *The topological fixed point property is a topological invariant.*

Proof: Suppose X has the *tfpp* and let $h : X \rightarrow Y$ be a homeomorphism. Suppose $f : Y \rightarrow Y$ is continuous. Then, $h^{-1} \circ f \circ h : X \rightarrow X$ is continuous and by supposition has a fixed point x. Then, $f(h(x)) = h(x)$ and Y has the *tfpp.*

Retraction: *A retraction of a set X onto a subset $Y \subseteq X$ is a continuous function $r : X \rightarrow f(X) = Y$ such that $f\,|_Y$ is the identity map.*

Remark 2**:** *The topological fixed point property is preserved under retractions.*

Proof: Suppose X has the *tfpp* and let $r : X \rightarrow Y$ be a retraction. Suppose $f : Y \rightarrow Y$ is continuous.Then, $f \circ r : X \rightarrow X$ is a continuous self–map of X, so for some x, $f \circ r(x) = x$. Since $x \in Y$, $r(x) = x$ and x is a fixed point of f. Thus, Y has the *tfpp.*

2.1.2 Background on Simplexes and Triangulations

The closed convex hull of a subset $A = \{a_{i,\dots,}a_k\} \subseteq \boldsymbol{R}^n$ is the set

$\overline{conv}(A) = \{\sum_{i=1}^{k} \lambda_i a_i \,;\, \lambda_i \geq 0, \sum \lambda_i = 1\}$. A subset S of $\boldsymbol{R}^n$ is called a **simplex** or **k-dimensional simplex**, provided there exists a set $V = \{v_0, \ldots, v_k\}$ such that the vectors $(v_1 - v_0), (v_2 - V_0), \ldots, (v_k - v_0)$ are linearly independent and $S = \overline{conv}\{v_0, \ldots, v_k\}$. The v_i are called the *vertices* of S. If context is clear, we may simply write $S = \{v_0, \ldots, v_k\}$ to refer to the simplex S. A ***p*–face** of S is the closed convex hull of any subset of p points in V. The ordering of the set V determines an orientation of the simplex. If two orderings differ by an even permutation, then they induce the same orientation. If they differ by an odd permutation, they induce opposite orientations of the simplex S. A finite collection K of simplexes that contains the faces of each of its members and is such that any two members of K who intersect do so in a face is called a **simplicial complex**. A **simplicial subdivision** of a k–simplex S is obtained by adding vertices to the vertex set of S and then adding new faces to S in such a way that a simplicial complex is obtained. A k–simplex in this new collection is called a ***k*–subsimplex** of the complex. A triangulation of a topological space X consists of a simplicial complex K and a homeomorphism $h : |K| \to X$, where $|K|$ is the complex K thought of as a subset of Euclidean space endowed with the subspace topology.

A **labeling** of a k–simplex or simplicial complex S is a function μ that maps the set of vertices $V = \{v_0, v_1, \ldots, v_k\}$ of S to the set of integers $\{0, \ldots, k\}$. A labeling of a k–simplex is said to be **proper** provided this mapping is a bijection. A simplicial subdivision S' of a k–simplex S is properly labeled provided $\mu|_S$ is a proper labeling and for any vertex $v \in S'$ contained in a face of S carrying the labels $i_0, i_1, \ldots, i_m$, $\mu(v)$ is one of $i_0, i_1, \ldots, i_m$. A k–subsimplex S of a properly labeled simplicial subdivision is called **distinguished** if μ maps S onto the set $\{0, 1, \ldots, k\}$. It is true that any properly labeled simplicial subdivision of a k–simplex S contains an odd number of distinguished k–subsimplexes.

2.1.3 Background in Analysis and Topology

Let $\boldsymbol{E}$ be a topological space and $X \subseteq \boldsymbol{E}$. A function f is **conti-**

nuous if the inverse image under f of an open set is open. An **open cover** of X is a collection of open sets whose union contains X. We say X is **compact** provided every open cover has a finite subcover. This is equivalent to stating that every collection of closed sets in X with the finite intersection property has nonempty intersection. A metric space is compact provided every sequence has a convergent subsequence. In $\boldsymbol{R}^n$, compactness is equivalent to closed and bounded. If f is continuous and X is compact, then $f(X)$ is compact. Any closed subset of a compact space is compact. If points in X can be separated by disjoint open sets, then we say X is **Hausdorff**. In a Hausdorff space, compact subsets are closed. A compact Hausdorff space is **normal** by which we mean closed sets can be separated by disjoint open sets.

A family $\{U_i\}_{i \in I}$ of subsets of X is called **neighbourhood finite** (nbd–finite) if each x in X has a neighbourhood V such that $V \cap U_i \neq \phi$ for at most finitely many $i \in I$. For X Hausdorff, a family $\{\beta_i\}_{i \in I}$ of continuous real valued maps is a **partition of unity** on X, provided the supports of the β_i form a nbd–finite closed covering of X, $0 \leq \beta_i(x) \leq 1$ and $\sum \beta i(x) = 1$, for each x in X. Given an open cover $\{U_i\}_{i \in I}$ of X, we say a partition of unity $\{\beta_i\}_{i \in I}$ is **subordinate** to this cover if for each i, the support of β_i lies in U_i. If X is (para)compact, then any open cover of X admits a partition of unity subordinate to it. A **topological vector space** is a vector space equipped with a topology such that scalar multiplication and vector addition are continuous. In what follows, we must be able to topologically distinguish points. Thus, we assume that all spaces are Hausdorff.

2.1.4 Upper Semicontinuous Multifunctions

Let X and Y be topological spaces. A **multifunction** $F : X \to 2^Y$ is a mapping that sends points x in X to subsets $F(x)$ of Y. We write, $\cup_{x \in X} F(x) = F(X)$. For $y \in Y$, we define the inverse of F at y to be $F^{-1}(y) = \{x \in X : y \in F(x)\}$. For $B \subseteq Y$, $F^{-1}(B) = \{x \in X : F(x) \cap B \neq \phi\}$. The **graph of** F is the set $Gr(F) = \{(x, y) \in X \times Y : y \in F(x)\}$. We say that F is **upper semicontinuous** (*usc*) at $x \in X$ if for every neighbourhood V of $F(x)$, there exists a neighbourhood U of x with $F(U) \subseteq V$. We say that $F : X \to 2^Y$ is *usc* if it is *usc* at every $x \in X$. If Y is compact and the images $F(x)$ are closed, then F is *usc* if and only if $Gr\,(F)$ is closed in $X \times Y$. In this case, if Y is compact, we also have that F is *usc* if $x_n \to x$, $y_n \to y$ and $y_n \in F(x_n)$ together imply that $y \in F(x)$. In a topological vector space, a *usc* multi-function with nonempty, compact, convex images is called a *cusco* for short. Some facts about *cuscos* are immediate.

***Proposition 1** : Let X be a compact subset of a topological vector space. If $F : X \to 2^X$ is a cusco and $f : X \to X$ is linear, then $f \circ F : f(X) \to 2^{f(X)}$ is a cusco.*

***Proposition 2** : Let $X \subseteq Y$ be subsets of a Banach space with X closed. If $F : X \to 2^X$ is a cusco and $f : Y \to X$ is continuous, then $F \circ f : Y \to 2^Y$ is a cusco.*

2.2 The Brouwer fixed point theorem

The Brouwer Fixed Point Theorem was first published in 1909 by Luitzen Egbertus Jan Brouwer (1881 - 1966) for continuous maps on $\boldsymbol{D}^3$. Then in 1910, for differentiable maps on $\boldsymbol{D}^n$ (the closed n-disk in $\boldsymbol{R}^n$). Finally, he proved the general version in 1912. Brouwer's fixed point theorem is the assertion that the class of compact convex sets in $\boldsymbol{R}^n$ has the fixed point property. As is often the case, Brouwer was not the first to prove "his" theorem. The result has its roots at least as far back as 1817 when Bolzano's Intermediate Value Theorem appeared which is equivalent to the Brouwer Fixed Point Theorem in dimension 1.

***Theorem** : Let $f : \boldsymbol{R}^n \to \boldsymbol{R}^n$ be continuou and suppose that $|x_i| \leq a_i$, for some prescribed set of reals $a_i > 0$ and $1 \leq i \leq n$. Further, suppose that on each face $x_i = a_i$, we have $f_i(x) > 0$ and for $x_i = -a_i$, we have $f_i(x) < 0$. Then, there exists x such that $f(x) = 0$.*

Miranda's name has been attached to this theorem because in 1941 he proved that it was in fact equivalent to Brouwer's theorem. This wasn't the only equivalent result to preclude Brouwer's publication of his theorem. In 1904, Bohl used Green's Theorem to prove that there could be no retraction of the *n*–cube onto its boundary. There does not seem to be evidence that Bohl made the short leap from this to deduce Brouwer's theorem. Following this, in 1909 Brouwer proved the theorem in $\boldsymbol{R}^3$. Then, in 1910, using Kronecker indices, Hadamard published a proof of the theorem for general *n* in the appendix of a book by Tannery. It wasn't until 1912 that Brouwer himself published his proof in $\boldsymbol{R}^n$. He used simplicial approximations and the degree of a map to prove his theorem in the setting of an *n*–simplex.

1883 : Bolzano–Poincare–Miranda theorem.
1904 : Bohl proves no retraction of *n*–cube onto its boundary.
1909 : Brouwer proves Brouwer's theorem in $\boldsymbol{R}^3$.
1910 : Hadamard proves Brouwer's theorem in $\boldsymbol{R}^n$.
1912 : Brouwer proves Brouwer's theorem in $\boldsymbol{R}^n$.

It seems to have been generally accepted that Brouwer knew the theorem to be true in 1910. In fact, Hadamard knew of the theorem through a letter from Brouwer himself which he received that same year. Still, it is clear that Brouwer was not the first to prove the Brouwer fixed point theorem. It is somewhat ironic that decades his name was the one attached to the result when decades later his intuittionist philosophy dictated he reject his non constructive proof. We state, now, Brouwer's fixed point theorem.

***Theorem* [*Brouwer's Fixed Point*]:** *Any compact convex subset of* $\boldsymbol{R}^n$ *has the fixed point property.*

Brouwer's original proof used complicated ideas such as the degree of a map. In the decades that followed, mathematicians searched for proofs that were simpler or somehow better or that used the language of a certain field. In this chapter, we explore various proofs of the theorem, ranging from the early 20th century upto the beginning of the 21st.

2.2.1 Proof by Non analytic Methods

In general, the proofs in this section are the earlier methods used to prove Brouwer's theorem. Brouwer himself used the notion of the degree of a map on the sphere and it is a proof based on this literature that we present first. The second method we illustrate uses the famous KKM theorem. It is relatively easy to derive Brouwer's theorem from KKM but the section has been shortened in that we leave out the proof of Sperner's lemma. Still, the proof of the lemma is not hard and thus this may be the most elementary proof we show. The final nonanalytic proof we give uses the nth homology groups of a topological space X. There is some difficulty in setting up the language in this section but, once, it is in place, it provides an immediate proof of Brouwer's theorem.

2.2.1 (a) The Degree of a Self–map on S^{n-1}

The proof given by Brouwer in1912 was based on the notion of the degree of a continuous function $f : \boldsymbol{S}^{n-1} \to \boldsymbol{S}^{n-1}$. The following version of the proof can be found in Dugundji. When $n = 2$, the degree of f, $\deg(f)$, can be thought of as the net number of times $f(x)$ travels around the unit circle as we let x make one counterclock-wise trip around. Formally, choose set of points $\{x_0, x_1, \dots, x_p\}$ taken in counterclockwise order around $\boldsymbol{S}^1$ such that $|x_{i+1} - x_i| < 1$. Then, each segment $[x_i, x_{i+1}]$ is a 1–simplex in $\boldsymbol{R}^2$. Think of this segment as an arc on the unit circle rather than a straight line segment. That is, let $[x_i, x_{i+1}]$ be the projection from the origin through the line segment onto to circle. Then, the union $\bigcup[x_i, x_{i+1}] = T$ is a triangulation of $\boldsymbol{S}^1$.

The set $\{f(x_0), \dots, f(x_p)\}$ will be another ordered set of points around $\boldsymbol{S}^1$ but the points may not follow each other counterclockwise around $\boldsymbol{S}^1$. The function f may reverse the orientation of some pairs of these points. We say that the image simplex $[f(x_i), f(x_{i+1})]$ has positive orientation if as x travels from x_i to x_{i+1}, $f(x)$ traverses the segment $[f(x_i), f(x_{i+1})]$ in a counterclockwise manner. If $f(x)$ travels in the opposite direction, we say this 1–simplex has negative orientation.

Next, fix $x \in \boldsymbol{S}^1$ such that x is not on the boundary of any of the image segments $[f(x_i), f(x_{i+1})]$. Then, the number of positively oriented image segments containing x minus the number of negatively oriented segments is called the degree of f at x with respect to the given triangulation T. With a little effort, it can be seen that this degree is actually independent of x and T.

In general, an n–simplex S in $\boldsymbol{R}^n$ is the convex hull of a set of $n + 1$ points called vertices. By fixing the order of the vertices, we consider S to be an ordered simplex. S is said to be nondegenerate provided the volume of this hull in $\boldsymbol{R}^n$ is nonzero. That is, S is nondegenerate provided its $n + 1$ vertices do not all lie on an $(n-1)$ – hyperplane. If we write $S = \overline{conv}\{x_0, \ldots, x_n\}$ and $x_i = (x_i^1, \ldots, x_i^n) \in \boldsymbol{R}^n$, then this nondegeneracy condition is equivalent to

$$\text{Det}\,(S) = \begin{vmatrix} x_0^1 & \cdots & x_0^n & 1 \\ x_1^1 & \cdots & x_1^n & 1 \\ \cdots & \cdots & \cdots & \cdots \\ x_n^1 & \cdots & x_n^n & 1 \end{vmatrix} \neq 0$$

Further, if $\det(S) > 0$ we say the n – simplex S is **positively oriented**. If $\det(S) < 0$, then S is **negatively oriented**. From the rules for matrix determinants, we see that even permutations of the order of the vertices of S will not change its orientation, odd ones will reverse the sign of $\det(S)$.

Lemma A : *Let S and S' be two oriented n–simplexes with*

$$S = \{x_0, x_1, \ldots, x_n\},$$
$$S' = \{x', x_1, \ldots, x_n\}.$$

Then, S and S' have the same orientation if and only if x_0 and x_0' lie on the same side of the $(n-1)$ – hyperplane H containing $\{x_1, \ldots, x_n\}$.

Proof : Let $S = \{x_0, x_1, \ldots, x_n\}$ and $S' = \{x_0', x_1, \ldots, m\, x_n\}$. If x_0 and

$x_0^{/}$ lie on opposite sides of H, then $\bar{x} = tx_0 + (1 - t)x_0^{/} \in H$ for some $t \in (0, 1)$. Then, the n–simplex $\bar{S}$ with vertices $\{\bar{x}, x_1, \ldots, x_n\}$ is degenerate and $\det(\bar{S}) = t \det(S) + (1 - t) \det(S^{/}) = 0$. This is possible only when S and $S^{/}$ have opposite orientation.

As in the discussion for $n = 1$, we shall need to look at triangulations living on the unit sphere $\mathbf{S}^{n-1}$. Any set of n points on $\mathbf{S}^{n-1}$ that do not lie on the same $(n-2)$ – hyperplane determine an $(n-1)$ – simplex S. We say a simplex S is proper if $\operatorname{diam}(S) < 1$. In this case, the projection from the origin through S onto $\mathbf{S}^{n-1}$ determines a set $\mathcal{S}$ that is proper in the same sense, i.e., $\operatorname{diam}(\mathcal{S}) < 1$. Such a projecttion will be called the spherical $(n-1)$ – simplex corresponding to S. The vertex set of $\mathcal{S}$ is the same as that of S. The ordering and orientation of the projection is inherited from the original simplex. A spherical $(n-1)$ – simplex $\mathcal{S}$ will be said to be degenerate in the case that the simplex determined by its vertices along with the point 0 is a degenerate n – simplex. A triangulation of $\mathbf{S}^{n-1}$ is a finite collection $T = \bigcup \mathcal{S}$ of nondegenerate ordered spherical $(n-1)$–simplexes covering $\mathbf{S}^{n-1}$ and satisfying two properties. First, members of T intersect only in a common face and second, for any $\mathcal{S} \in T$, each $(n-2)$ –face of $\mathcal{S}$ is shared with exactly one other member of T. The vertices of T are the union of the vertices of the $\mathcal{S} \in T$. A function that maps vertices of T into $\mathbf{S}^{n-1}$ is called a proper vertex map provided that for each $\mathcal{S} \in T$, the simplex determined by the image of the vertices of $\mathcal{S}$ is proper and hence has an associated proper spherical simplex.

With the above definitions in place, let T be a triangulation of $\mathbf{S}^{n-1}$. Using the orientability of $\mathbf{S}^{n-1}$, assume all simplexes are oriented positively. Let $f : T \to \mathbf{S}^{n-1}$ be a proper vertex map. Fix $x \in \mathbf{S}^{n-1}$ such that x is not on the boundary of $f(\mathcal{S})$ for any $\mathcal{S} \in T$. Let $p(f, T, x)$ be the number of positively oriented spherical simplexes $f(\mathcal{S})$ containing x and $n(f, T, x)$ the number of negatively oriented spherical simplexes containing x. Then, we define the degree of f with respect to T and x to be

$$\deg(f, T, x) = p(f, T, x) - n(f, T, x).$$

Lemma B *: For a given triangulation T of* $\mathbf{S}^{n-1}$*, with each* $\mathcal{S} \in T$ *positively oriented and a proper vertex map f,* deg(*f, T, x*) *is independent of the choice of x.*

Proof: We prove the case where for each $\mathcal{S} \in T, f(\mathcal{S})$ determines a nondegenerate spherical $(n-1)$ – simplex. Let $f : T \rightarrow \mathbf{S}^{n-1}$ be a proper vertex map. Pick any two points $y, z \in \mathbf{S}^{n-1}$ that do not lie on the boundary of $f(\mathcal{S})$ for any $\mathcal{S} \in T$. We will show that deg(f, T, y) = deg(f, T, z). To this end, let C be an arc in $\mathbf{S}^{n-1}$ joining y and z that doesn't pass through any face of dimension less than $(n-2)$ of any spherical $(n-1)$ – simplex of $f(\mathcal{S})$. We consider what happens to deg(f, T, x) as we let x move from y to z.

Clearly, for the degree to change, x has to travel across some $(n-2)$ – face of some simplex in $f(T)$. Call one such simplex $f(\mathcal{S}_1)$ where $\mathcal{S}_1 = \{x_0, x_1 \dots , x_{n-1}\}$, and the $(n-2)$ –face x crosses is $A = \{f(x_1), f(x_2), \dots, f(x_{n-1})\}$. Now, to $\mathcal{S}_1$ there corresponds exactly one other simplex in T that shares the face $\{x_1, \dots , x_{n-1}\}$. Call this simplex $\mathcal{S}_2$ and write $\mathcal{S}_2 = \{x_0', x_2, x_1, \dots , x_{n-1}\}$, where the ordering of the vertices is chosen to give $\mathcal{S}_2$ positive orientation. Note that $f(\mathcal{S}_2)$ is a simplex in $f(T)$ that shares the $(n-2)$ – face that our point x crosses. We write

$$f(\mathcal{S}_1) = \{f(x_0), f(x_1), f(x_2), \dots , f(x_{n-1})\}$$
$$f(\mathcal{S}_2) = \{f(x_0'), f(x_2), f(x_1), \dots , f(x_{n-1})\}.$$

There are two cases to consider. First, suppose $f(x_0)$ and $f(x_0')$ are on the same side of the hyperplane determined by the vertices of A. As x travels through the face A it either enters both $f(\mathcal{S}_1)$ and $f(\mathcal{S}_2)$, or it leaves both. Then by *lemma A*, the simplexes $f(\mathcal{S}_1)$ and $f(\mathcal{S}_2)$ have opposite orientation. In either case, the net change in $p(f, T, x) - n(f, T, x)$ is zero.

Next, suppose $f(x_0)$ and $f(x_0')$ lie on opposite sides of A. As x passes through A it must leave one of $f(\sigma_1)$ or $f(\sigma_2)$ and enter the other. Again, by the lemma, $f(\sigma_1)$ and $f(\sigma_{-2})$ have the same orien tation. Again, there is no net change in deg(f, T, x).

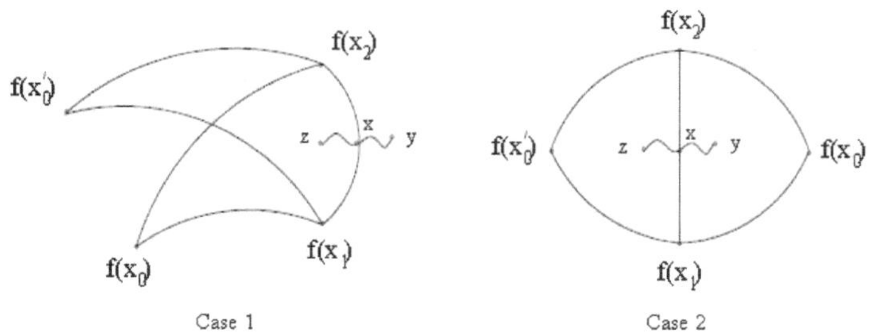

Figure: The two cases in the proof of *lemma B*

We address the case when some of the $f(\sigma)$, $\sigma \in T$ may be degenerate. To do this, we will reduce it to the previous one by approximating f with another proper vertex map g. Suppose f maps σ to a degenerate simplex $f(\sigma)$. This means that some vertex in σ gets mapped to the $(n-3)$ –hyperplane containing the remaining $(n-2)$ vertices. Now, y and z lie in the interior of $f(\sigma)$ for each σ. Thus, we can slightly perturb f at the offending vertex so that its image is no longer in the $(n-1)$ –hyperplane and we haven't altered the orientation or number of simplexes $f(\sigma)$ containing y or z. That is, we can find an $\varepsilon > 0$ and a proper vertex map $g : T \rightarrow S^{n-1}$ that has no degenerate $g(\sigma)$ and $|f(x) - g(x)| < \varepsilon$ for each vertex $x \in T$. We have that deg(f, T, y) = deg(g, T, y) and deg(f, T, z) = deg(g, T, z). So, via g, and the previous nondegenerate case, we see deg(f, T, y) = deg(f, T, z).

Thus, we may write deg(f, T) without any ambiguity. From the above argument, we obtain the following useful lemma.

Lemma C: *Let f, $g : T \rightarrow S^{n-1}$ be proper vertex maps and $x \in S^{n-1}$ not on the boundary of any of any $f(\sigma)$, $\sigma \in T$. There exists an $\varepsilon > 0$ such that if $|f(y) - g(y)| < \varepsilon$ for all vertices in T, then* deg(f, T, x) = deg(g, T, x).

Given a triangulation T, we can refine T by adding a finite set of points to its vertex set. We, then, need only add faces to T to preserve the triangulation properties. Barycentric subdivision is one example of a refinement procedure.

Until now, the function f has been defined only on the vertices of T. We can extend the above concepts to a continuous $f : \mathbf{S}^{n-1} \to \mathbf{S}^{n-1}$ since any such function induces a vertex map on a T by restricting its domain to the set of vertices of T. For an arbitrary triangulation T, f may not be a proper vertex map. It follows from the continuity of f and the compactness of $\mathbf{S}^{n-1}$ that we can find a refinement T' of T such that diam($f(S)$) $<$ 1 for each $S \in T'$. Thus, we can speak of the degree of a continuous function f with respect to a triangulation T.

Lemma D: deg(f, T) *is independent of the choice of* T.

Proof: Let T_1 and T_2 be two triangulations of $\mathbf{S}^{n-1}$, both inducing proper vertex maps of f. Then, let T_3 be a common refinement of both. We will show that if T' is a refinement of T, then deg(f, T) = deg(f, T').

Suppose T' is a refinement of T formed by adding a single vertex to T. Note that if the image of just the new simplexes created by adding this new point covers all of $\mathbf{S}^{n-1}$, then there must have been a simplex in the original triangulation that violated the property diam($f(S)$) $<$ 1. Thus, we can pick $x \in \mathbf{S}^{n-1}$ that is not in the image of any of the newly created simplexes. Thus, deg(f, T, x) = deg(f, T', x). By induction, we see that any refinement of T will not alter the degree. Thus, via the common refinement, we see

$$\deg(f, T_1) = \deg(f, T_2).$$

From this point, we may write deg(f) to refer to the degree of f.

Theorem E: *If* f, $g : \mathbf{S}^{n-1} \to \mathbf{S}^{n-1}$ *are homotopic, then* deg(f) = deg(g).

Proof: Let f, $g : S^{n-1} \to S^{n-1}$ and $F : S^{n-1} \times I \to S^{n-1}$ be a homotopy of f and g. So, F is continuous and $F(x, 0) = f(x)$, $F(x, 1) = g(x)$. First, note that by compactness of $S^{n-1} \times I$, F is uniformly continuous. Thus, $\exists \delta > 0$ such that for any $t \in I$ and $x, y \in S^{n-1}$ satisfying $|x - y| < \delta$, we have $|F(x, t) - F(y, t)| < 1$. Choose a triangulation T of S^{n-1} such that $\text{diam}(S) < \delta$ for all $S \in T$. Then, for each t, the function $F(\cdot, t)$ induces a proper vertex map of T.

Now, by the previous lemma, there is an $\varepsilon > 0$ such that if $|F(x, t) - h(x)| < \varepsilon$ for every vertex $x \in T$, then $\deg(F(\cdot, t)) = \deg(h)$. Fix such an ε. Again, by uniform continuity of F, choose $\delta > 0$ such that for any $x \in S^{n-1}$, $|F(x, t) - F(x, t')| < \varepsilon$ whenever $|t - t'| < \delta$. Thus, we have shown that the function mapping t to $\deg(F(\cdot, t))$ is a continuous integer valued function and hence, must be a constant function. Then, $\deg(f) = \deg(g)$.

Example : *The degree of* $id : S^n \to S^n$ *is* 1.

Proof: Follows from the definition.

With the above machinery in place, we can prove that the unit sphere is not a retract of the unit ball. This result is a well known equivalence of Brouwer's theorem.

Lemma F: *Suppose* $f : S^{n-1} \to S^{n-1}$ *has a continuous extension to* B^n. *Then, f is homotopic to a constant fun-c tion.*

Proof: Let $\overline{f} : B^n \to S^{n-1}$ be a continuous extension of f. Define $F : S^{n-1} \times I \to S^{n-1}$ by $F(x, t) = \overline{f}((1 - t)x)$. Then, F is a homotopy of f to the constant map that sends x to $\overline{f}(0)$.

Theorem G: S^{n-1} *is not a retract of* B^n.

Proof: Suppose $r : B^n \to S^{n-1}$ is a retraction. Then, $r|_{S^{n-1}} = id : S^{n-1} \to S^{n-1}$ would have a continuous extension to B^n. Thus, by the previous lemma, $id : S^{n-1} \to S^{n-1}$ is homotopic to a constant function. On the other hand, the degree of the constant map is 0 while the degree of $id : S^{n-1} \to S^{n-1}$ is 1. By *theorem E*, these functions cannot be homotopic.

From here, we easily deduce Brouwer's theorem. Assume the continuous map $f : B^n \to B^n$ is fixed point free. Then, define g to be the function that maps a point x to the projection from 0 through x onto S^{n-1}. Then, g is a continuous retract of the unit ball onto the unit sphere violating the above theorem.

2.2.1 (b) The KKM Theorem

The following section looks at a proof of Brouwer's theorem that follows from an elegant combinatorial result on simplicial subdivisions proven in 1928 by Sperner. A year after this lemma was published, Knaster, Kuratowski and Mazurkiewicz used it to prove the so called *KKM* theorem from which they deduced the Brouwer fixed point theorem. The ease at which they obtained Brouwer's theorem caused speculation as to the possible equivalence of these three theorems. The question remained open for almost 50 years until in 1974 Yoseloff showed that Brouwer implies Sperner's lemma. Below, we show how to obtain the *KKM* theorem from Sperner's lemma and then, apply it to obtain Brouwer's theorem.

***Theorem H* [Sperner's Lemma] :** *Any properly labeled simplicial subdivision of a k– simplex has an odd number of distinguished k–subsimplexes.*

.

***Theorem I* [KKM]:** *Let* $S = \overline{conv}\{v_0, v_1, \ldots, v_k\}$ *be a k–simplex. Suppose* $A_0, A_1, \ldots, A_k$ *are closed subsets of S such that* $\overline{conv}\{v_{i0}, v_{i1}, \ldots, v_{im}\} \subseteq \bigcup_{j=0}^{m} A_{ij}$ *holds for any subset* $\{v_{ij}\}$ *of* $\{v_i\}^k_{i=0}$. *Then,*

$$\cap_{i=0}^{k} A_i \neq \phi .$$

Proof: Let $S = \overline{conv}\{v_0, v_1, \ldots, v_k\}$ be a k–simplex. For each $n \in N$, there is a simplicial subdivision S^n of S such that the diameter of each k–subsimplex of S^n is less than $1/n$. Dress S with a proper labeling of its vertices, so v_i is labeled with i. We extend this to a proper labeling of S^n as follows. For each vertex q of S^n, there is a smallest face $\overline{conv}\{v_{io}, v_{i1}, \ldots, v_{im}\}$ of S containing q. Then, by our assumption, $q \in A_{ij}$ for some j, $0 \le j \le m$. Assign to q the label ij . Thus, we obtain the desired labeling on S^n.

Now, by Sperner's lemma, S^n has a distinguished k–subsimplex, say $\overline{conv}\{q^n{}_o, q^n{}_1, \ldots, q^n{}_k\}$.

Upon relabeling if necessary, we can assume that the vertex $q^n{}_i$ carries the label i and thus is a member of A_i by our construction. Now, compactness of S guarantees a convergent subsequence of $\{q^n{}_i\}^\infty{}_{n=1}$ for each i. Since the diameter of the k–subsimplexes goes to zero, these sequences must converge to a common point. Since each A_i is closed in S, this limit must be in $\bigcap_{i=0}^{k} A_i$.

From the above lemma, we can easily deduce Brouwer's theorem. Let S be a k–simplex with vertices $\{v_0, v_1, \ldots, v_k\}$, and f be a continuous self–map on S. For $x \in S$, we have, $x = \sum_{i=0}^{k} x_i v_i$ with $\sum_{i=0}^{k} x_i = 1$ and $x_i \ge 0$. Define $A_i = \{x \in S : f_i(x) \le x_i\}$.

From the continuity of the components of f, we see that each A_i is indeed a closed subset of S. By applying the *KKM* theorem to these A_i, we get some point $y \in S$ such that $f_i(y) \le y_i$ for each i. But, since $\sum y_i = 1 = \sum (f(y))_i$, we must have $y_i = (f(y))_i$ for each i. That is, $f(y) = y$.

Of course, this theorem extends to any closed convex subset in $\boldsymbol{R}^n$ since any such a set is a retract of an n–simplex.

2.2.1 (c) Via Homology Groups

To give a good sampling of the methods with which Brouwer's theorem may be proven, it is necessary to discuss homology groups. Of all the methods discussed, it is from the language of the material in this section that Brouwer's theorem flows most naturally. The difficulty is that this language takes quite a bit of work to establish. For this reason, we will show how homology groups are defined on simplicial complexes and compute some of these groups for triangularions of the ball and sphere. We will discuss, though not prove, how these ideas are extended to arbitrary topological spaces and use the homology groups of $\boldsymbol{B}^n$ and S^{n-1} to prove Brouwer's theorem.

The first thing we must do is define $H_q(K)$, the qth–homology group of K, where K is a simplicial complex. To do so, we consider the set of all q–simplexes in K. Each q– simplex can be oriented in one of two ways. To each q–simplex z in K, we assign one of those orientations to be positive and call $-z$ the q–simplex z oriented in the opposite way. The qth–chain group of K, denoted $C_q(K)$, is the free abelian group generated by these oriented q–simplexes. The elements of $C_q(K)$ are called q–chains. The boundary function $\partial : C_q(K) \to C_{q-1}(K)$ maps a q–chain to its $(q-1)$–dimensional boundary, a $(q-1)$–chain in $C_{q-1}(K)$. It is defined for each q–simplex $z = (v_0, \dots, v_q)$ by

$$\partial(v_0, \dots, v_q) = \sum_{i=0}^{q} (-1)^i (v_0, \dots, \hat{v}_i, \dots, v_q),$$

where$(v_0, \dots, \hat{v}_i, \dots, v_q)$, is the simplex formed by deleting v_i from the vertex set of z. Then, ∂ is extended linearly for longer q–chains.

Let $Z_q(K)$ be the kernel of $\partial : C_q(K) \to C_{q-1}(K)$. Thus, $Z_q(K)$ consists of those q– chains that have zero boundary. We call these chains q–cycles. Next, we define $B_q(K)$ to be the image of the map $\partial : C_{q+1}(K) \to C_q(K)$. Elements of $B_q(K)$ are called bounding q–cycles as they bound the $(q+1)$–simplex that was their preimage under ∂.

Proposition 3: $B_q(K) \subseteq Z_q(K)$.

Proposition 4: *Let K be the simplicial complex formed by an n–simplex with all of its faces. Then, $H_q(K) = 0$ for $0 < q \leq n$.*

Proposition 5: *For* $0 \leq q \leq n-1$, $H_q(K^{n-1}) = 0$.

Proposition 6: $H_{n-1}(K^{n-1}) = \mathbf{Z}$

Theorem J: *A continuous function* $f : |K| \rightarrow |L|$ *induces a homomorphism* $f_{q*} : H_q(K) \rightarrow H_q(L)$ *for each q.*

Theorem K: *If* $: |K| \rightarrow |K|$ *is the identity, then* $f_{q*} : H_q(K) \rightarrow H_q(K)$ *is the identity homomorphism. Also, if* $f : |K| \rightarrow |L|$, $g : |L| \rightarrow |M|$ *are continuous, then* $(g \circ f)_{q*} = g_{q*} \circ f_{q*}$.

We give a brief discussion of the proof of *Theorem J*. *Theorem J* is proven first for a class of mappings called simplicial maps that take simplexes of K linearly into simplexes of L.

Given $s : |K| \rightarrow |L|$, we define $s_q : C_q(K) \rightarrow C_q(L)$ naturally by $s_q(v_0, \ldots, v_q) = (s(v_0), \ldots, s(v_q))$, when each $s(v_i)$ is distinct, and zero otherwise. It is then shown that s_q sends q–cycles and bounding q–cycles in $C_q(K)$ into respectively q–cycles and bounding q–cycles in $C_q(L)$. Thus, s_q induces a homomorphism $s_{q*} : H_q(K) \rightarrow H_q(L)$. To extend this idea to any continuous function $f : |K| \rightarrow |L|$, it is shown that we can subdivide K and L into finer simplicial complexes for which we can find a simplicial map s that is as close as we want to f.

With the above machinery, we can prove Brouwer's theorem. Suppose $f : \boldsymbol{B}^n \rightarrow \boldsymbol{B}^n$ is continuous and fixed point free. Then, as is familiar by now, define $g : \boldsymbol{B}^n \rightarrow \boldsymbol{S}^{n-1}$ by mapping x to the intersection of the ray extending from $f(x)$ through x with $\boldsymbol{S}^{n-1}$. Then, g is a continuous function. Let $\imath : \boldsymbol{S}^{n-1} \rightarrow \boldsymbol{B}^n$ be the inclusion mapping $\imath(x) = x$. Then, by *Theorem J*, both g and $\imath$ induce homomorphisms g_* and $\imath_*$ of the $(n-1)$th homology groups

$$H_{n-1}(\boldsymbol{S}^{n-1}) \xrightarrow{\imath_*} H_{n-1}(\boldsymbol{B}^n) \xrightarrow{g_*} H_{n-1}(\boldsymbol{S}^{n-1})$$

Now, since $g \circ \imath : \boldsymbol{S}^{n-1} \rightarrow \boldsymbol{S}^{n-1}$ is the identity, by *Theorem K*, $g_* \circ \imath_*$ is the identity homomorphism. Then, g_* must be onto. But, from *Proposition 4*, $H_{n-1}(\boldsymbol{B}^n) = 0$ while *Proposition 6* tells us $H_{n-1}(\boldsymbol{S}^{n-1})$

= $\boldsymbol{Z}$. Thus, we have a contradiction. So, f must have had a fixed point.

2.2.2 *Proof by Analytic Methods*

From ***section 2.2.1 (a)***, we could have proceeded to define the characteristic function of a non vanishing vector field $f : \boldsymbol{S}^{n-1} \rightarrow \boldsymbol{S}^{n-1}$. Then, using facts about the degree of this function, we would have obtained another classical result in topolog, known as the Hairy ball theorem. Until the 1970's, both this result and Brouwer's theorem were proven using combinatorial arguments, homology theory, differential forms, or geometric topology. In 1978, John Milnor published self described "strange" proofs of these results that are nicely analytic in nature.

2.2.2 *(a) The Hairy Ball Theorem*

Milnor's proof is interesting not only because of its analytic nature, but, because it follows from a calculation of volume in $\boldsymbol{R}^n$ and the fact that $(1+t^2)^{n/2}$ is not a polynomial for odd n.

Lemma (i) : *Let $f : A \rightarrow \boldsymbol{R}^n$ be continuously differentiable over a neighborhood of the compact set A. Then, there exists a Lipschitz constant L such that for all $x, y \in A$, $||f(x) - f(y)|| \le L||x - y||$.*

Proof: Cover A with a finite number of balls $U_1, U_2, \ldots, U_p$ such that f is continuously differenttiable on $\overline{U_k}$, $1 \le k \le p$. First, we obtain a Lipschitz constant for f on U_k. By continuity of the partials on $\overline{U_k}$, we can choose a constant $c^k{}_{ij} = \max\limits_{i,j,k} \left\{ \frac{\partial f_i}{\partial f_j}(x) \ : \ x \in \overline{U_k} \right\}$. Using the triangle inequality, and the Mean Value property we obtain for $x, y \in U_k$,

$$||f(x) - f(y)|| \le \sum_{i=1}^{n} | f_i(x) - f_i(y) | \le \sum_{i=1}^{n} \sum_{j=1}^{n} c_{ij}^{k} \|x - y\|$$

$$= L_k \, || \, x\text{-}y \, || \quad \text{where } L_k = \sum_{i,j=1}^{n} c_{ij}^{k}.$$

Next, we consider the set of $x, y \in A$ such that x and y are not both in one of the U_k. This set can be expressed as $W = (A \times A) \setminus \bigcup_{k=1}^{p} U_k \times U_k$. Consider the function $g : W \to \boldsymbol{R}^n$ defined by $g(x, y) = || x-y ||$. This function is continuous over the compact set W, and thus, achieves its minimum. Since $x \neq y$, we have $\min_{y\ y \in W} ||x - y|| = \varepsilon > 0$. So, we obtain the following bound:

$$|| f(x) - f(y) || \leq \varepsilon^{-1} \operatorname{diam} f(A) \, || x - y ||.$$

Choose $L = \max\{L_1, L_2, \ldots, L_p, \varepsilon^{-1} \operatorname{diam} f(A)\}$ and we have obtained the desired Lipschitz constant.

Lemma (ii): *Suppose $A \subseteq \mathbf{R}^n$ is compact and $v : A \to \mathbf{R}^n$ is continuously differentiable in a neighbourhood of A. Then, there exists an interval $(-\varepsilon, \varepsilon)$ on which the function $t \mapsto |f_t(A)|$ is a polynomial.*

Proof: We will apply the Change of Variables formula to obtain the desired expression for the volume of $f_t(A)$. Thus, we need to show that f_t is one–to–one and continuously differentiable and that $D_x f_t$ (the derivative of f_t at x) is invertible for $x \in A$. To this end, by the previous lemma, let L be the Lipschitz constant for v and suppose $|t| < L^{-1}$.Then, $f_t(x)-f_t(y)$ implies that$||v(x)-v(y)||= t^{-1}||x-y|| \leq L|| x-y ||$. From out choice of t, this is possible only if $x = y$. Thus, f_t is one–to–one. Continuous differentiability of f_t follows from that of its component parts.

Next, $D_x f_t = I + t \left[\frac{\partial v_i}{\partial x_j}(x) \right]$ has a strictly positive determinant for t sufficiently small, say less than K. Set $\varepsilon = \min\{K, L^{-1}\}$. The for $|t| < \varepsilon$ and $x \in A$, we have $D_x f_t$ is invertible. Thus, we may express the volume of $f_t(A)$ by $\operatorname{vol} f_t(A) = \int_A |\det(D_x f_t)| \, dx$.

We may write det $(D_x f_t) = 1+ta_1(x) + t^2 a_2(x) +\ldots+ t^n a_n(x)$, where each a_i is a continuous function. Upon integrating this expression over A, we obtain

$$\text{vol}\, f_t(A) = \text{vol}A + t\alpha_1 + t^2\alpha_2 +\ldots + t^n\alpha_n,$$

where $\alpha_i = \int_A a_i(x)\,dx$.

Lemma (iii) *: Suppose* $v : S^{n-1} \to R^n$ *is a normed vector field tangent to* S^{n-1} *continuously differentiable on a neighbourhood of* S^{n-1}*. Then, for* $t > 0$ *small enough, the function* $f_t : S^{n-1} \to (1 + t^2)^{½} S^{n-1}$ *is onto.* (Contraction mapping and Banach Contraction Principle used in this lemma are discussed in Chapter 3)

Proof: First, we show the function is well defined. Indeed, for $x \in S^{n-1}$, the norm of $ft(x)$ can be computed by considering $\| f_t(x) \|^2 = \|x + tv(x)\|^2 = \langle x + tv(x), x + tv(x)\rangle$. From this, we obtain $\| f_t(x)\| = \sqrt{(1 + t^2)}$.

To show f_t is onto, first define A to be the set $A = \{x \in R^n : 1/2 \le \|x\|\ 3/2\}$ and extend f_t to all of A by setting $v(x) = \|x\| v\left(\frac{x}{\| x \|}\right)$ for x no on the unit sphere. Note that v is still continuously differentiable and so by *Lemma (i)* , has a Lipschitz constant L on A.

Fix $w \in \sqrt{(1 + t^2)}S^{n-1}$ and let $z \in S^{n-1}$ be such that $w = \sqrt{(1 + t^2 z)}$. Then, for $t < \min\{1/3, L^{-1}\}$, the function $g : x \mapsto z - tv(x)$ maps A into A and is a contraction mapping. Thus, by the Banach Contraction Principle, g has a fixed point, say x, in A. So, $x = z - tv(x)$. Since $z \in S^{n-1}$, we have $\| x + tv(x)\|^2 = 1$. Expanding the inner product $\langle x+tv(x), x+tv(x)\rangle$ gives $\|x\| = (1+t^2)^{1/2}$. Then, $y = (1+t^2)^{1/2} x \in S^{n-1}$, and $y+tv(y) = w$. Thus, f_t is onto.

Theorem (iv) [Hairy Ball Theorem - Weak Version]: *The sphere* S^{2k} *does not possess a continuously differentiable field of unit tangent vectors.*

Proof: Suppose the contrary and let $v : \mathbf{S}^{2k} \to \mathbf{R}^n$ be such a vector field. Select $0 < a < 1 < b$ and extend v to the set $A = \{x \in \boldsymbol{R}^n : a \leq \|x\| \leq b\}$ by defining $v(x)=\|x\|v\left(\frac{x}{\|x\|}\right)$ as in *Lemma (iii).* Note that for $x \in \boldsymbol{S}^{n-1}$, and $r > 0$, $F_t(rx) = rx + tv(rx) = rf_t(x)$. Thus, by *Lemma(iii),* f_t maps the sphere S_r of radius r onto $(1 + t^2)^{1/2} S_r$. Thus, $f_t(A) = (1+t^2)^{½} A$ for small enough t and so

$$|f_t(A)| = (1+t^2)^{2k+1}\, 2|A|.$$

This contradicts *Lemma (ii)* since the right hand side $\sqrt{1+t^2}\,(1 + t^2)^k|A|$ cannot be a polynomial.

Theorem(v)**[Hairy Ball Theorem - Strong Version]:** *There does not exist a continuous non zero vector field v tangent to* $\mathbf{S}^{2k}$.

Proof: Suppose $v : \boldsymbol{S}^{2k} \to \boldsymbol{R}^n$ is such a vector field. Let $m = \min\{\|v(x)\| : x \in \boldsymbol{S}^{2k}\} > 0$. By the Weierstrass approxi- mation theorem, each component v_i of v can be approximated by a polynomial $p_i : \boldsymbol{S}^{2k} \to \boldsymbol{R}$ such that $\| p(x) - v(x) \| < m/2$ for all $x \in \boldsymbol{S}^{2k}$. Then, p is C^1 and non zero since $\|p(x)\| \geq \| v(x)\| - \|p(x) - v(x)\| \geq m - m/2 = m/2$. Using p, we obtain a C^1 nonzero vector field q that is tangent to $\boldsymbol{S}^{2k}$. Define $q(x) = p(x) - \langle p(x), x\rangle x$. Then, q inherits its continuous differentiability from p and

$$\begin{aligned}
\|q(x)\| &\geq \|p(x)\| - \|q(x) - p(x)\| \\
&> m/2 - |\langle p(x), x\rangle| \\
&= m/2 - |\langle p(x) - v(x), x \rangle| \\
&\geq m/2 - \|p(x) - v(x)_\| \\
&> 0.
\end{aligned}$$

The proof is complete upon observing that $\frac{q(.)}{\| q(.) \|}$ contradicts *Theorem (iv).*

To proceed to the proof of Brouwer's Theorem, we need one more piece of machinery. The unit ball in $\boldsymbol{R}^n$ can be projected onto the lower hemisphere of the unit sphere in $\boldsymbol{R}^{n+1}$ with a stereographic projection. Write $\boldsymbol{R}^{n+1} = \{(x, x_{n+1}) : x \in \boldsymbol{R}^n, x_{n+1} \in \boldsymbol{R}\}$. Then, for $x \in \boldsymbol{R}^n$ this projection is defined by

$$S_+(x) = \left(\frac{2x}{\|x\|^2+1}, \frac{\|x\|^2-1}{\|x\|^2+1} \right)$$

Hence, $S_+(x)$ is the intersection with $\boldsymbol{S}^n$ of the ray starting at $(0,1) \in \boldsymbol{S}^n$ and passing through $x \in \boldsymbol{B}^n$. Clearly, this mapping is C^1. The image of $\boldsymbol{B}^n$ under S_+ is the lower hemisphere of $\boldsymbol{S}^n$ which we denote by $S^n{}_-$.

Similarly, we can define the projection S_- from $(0,-1)$ of $\boldsymbol{B}^n$ onto $S^n{}_+$, the upper hemisphere of $\mathbf{S}^n$

$$S\_(x) = \left(\frac{2x}{\|x\|^2+1}, \frac{1-\|x\|^2}{1+\|x\|^2} \right)$$

Now we are ready to prove Brouwer's theorem.

Let $f : \boldsymbol{B}^{2k} \to \boldsymbol{B}^{2k}$ be a continuous function and suppose f leaves no point fixed. We will use f to define a nonzero C^1 vector field on $\boldsymbol{B}^{2k}$ that points directly outward at points on $\boldsymbol{S}^{2k-1}$.

Suppose F is such a vector field. For each $x \in \boldsymbol{B}^n$, $\{x + tF(x) : 0 \le t \le 1\}$ will be a segment in $\boldsymbol{R}^n$. The image of this set under S_+ will be an arc on $\boldsymbol{S}^n$ with initial point $S_+(x)$ lying in $\boldsymbol{S}^n{}_-$. On $S^{2k}{}_-$, we can define a continuous nonzero field of tangent vectors by setting

$$T-(y) = \frac{d}{dt}\, S_+(x + tF(x))|_{t=0}$$

for each $y \in \boldsymbol{S}^{2k}$, where $y = S_+(x)$. $T_-(y)$ is the tangent of the arc $S_+(x{+}tF(x) : 0 \le t \le 1)$. Note that since F points outward on S^{2k-1}, the projections of the segment $\{x{+}tF(x) : 0 \le t \le 1\}$ will be "vertical" and so the tangent of the projected arc will be $(\mathbf{0}, 1)$.

Similarly, we can define $T_+(y) = \frac{d}{dt} S_-(x + tF(x))|_{t=0}$ for $y \in S_+^{2k}$. Again, for y on the equator, $T_+(y) = (0, 1)$.

Now, we define $T : \boldsymbol{S}^{2k} \to \boldsymbol{R}^{n+1}$ by

$$T(y) = \begin{cases} T_-(y) & for\ y \in S_-^{2k} \\ T_+(y) & for\ y \in S_+^{2k} \end{cases}.$$

Then, T is a continuous nonvanishing vector field tangent to $\boldsymbol{S}^{2k}$. This contradicts the Hairy ball theorem.

The required field $F : \boldsymbol{B}^{2k} \to \boldsymbol{R}^n$ can be defined by

$$F(x) = x - \left(\frac{1 - \langle x, x \rangle}{1 - \langle x, f(x) \rangle} \right) f(x)$$

Clearly, F is C^1 and points outward for $x \in \boldsymbol{S}^{2k-1}$. To see that F is nonzero, suppose the contrary. Then, for some x, f_x is a scalar multiple of x and thus $\langle x, f(x) \rangle x = \langle x, x \rangle f(x)$. It follows from the definition of F that $x = f(x)$ contradicting the hypothesis that f has no fixed points.

To complete the proof, suppose $f : \boldsymbol{B}^{2k-1} \to \boldsymbol{B}^{2k-1}$ is conti- nuous and fixed point free. Then, so is the function $g : \boldsymbol{B}^{2k} \to \boldsymbol{B}^{2k}$ defined by $(x, x_{2k}) \mapsto (f(x), 0)$ contradicting the above.

2.2.2(b) Garcia's Proof

The following proof was found in a text on mathematical economics. In the text, the author relates an anecdote in which he complains to a colleague that there is no simple proof of Brouwer's theorem. The colleague, Adriano Garcia, responded by giving the following proof, based on Green's theorem. Much of the work done in this section is needed to define a rather tedious function which Garcia refers to as the discriminant.

The Existence of the Discriminant

We need the discriminant function δ to be real valued and to act on the set of continuous functions $f : \boldsymbol{R}^n \rightarrow \boldsymbol{R}^n$. To begin, let A be a neighbourhood of $\boldsymbol{S}^{n-1}$ and let Z be the set of twice continuously differentiable functions $f : A \rightarrow \boldsymbol{R}^n$ such that $||f(x)|| = 1$ on $\boldsymbol{S}^{n-1}$. Recall that the Jacobian of a function f is the matrix of partial derivatives, the ith row being the derivative of the ith component of f. Next, let $Ji(f)$ denote the Jacobian matrix of f with the ith column replaced with $f = (f_1 \ldots, f_n)$ and define the operator G on the set Z by

$$G(f) = (\det J_1(f), \det J_2(f), \ldots, \det J_n(f)).$$

Now, on Z, we define the real valued function δ by

$$\int_{S^{n-1}} \delta(f) = G(f) \cdot \mathbf{n} \, dS.$$

The function δ has the following properties.

Property (vi): *Let $f \in Z$ and suppose f has a normed continuously twice differentiable extension to all of $\boldsymbol{B}^n$. Then, $\delta(f) = 0$.*

Property (vii): *Let $f, g \in Z$. If $f(x) + g(x) \neq 0$ for all $x \in \boldsymbol{S}^{n-1}$, then $\delta(f) = \delta(g)$.*

To prove these facts we will have to know more about the behavior of the operator G.

Lemma (viii): *Let $f : \boldsymbol{R}^n \rightarrow \boldsymbol{B}^n$ be twice continuously differentiable in a neighborhood of $\boldsymbol{S}^{n-1}$. Then, $\nabla \cdot G(f) = n \det J(f)$.*

Proof: Recall that $\partial_k(det J_k) = \sum_{m=1}^{n} \det(J_{km})$, where J_{km} is the matrix J_k with ∂_k applied to its mth column. Note since $\partial_k \partial_m f = \partial_m \partial_k f$, we see that J_{km} and J_{mk} will be the same but with columns m and k interchanged. Thus, $\det J_{km} = -\det J_{mk}$. Also, it is clear that $J_{kk} = J(f)$. Then, we obtain

$$
\begin{aligned}
\nabla \cdot (G(f)) &= \sum_{k=1}^{n} \partial_k (\det J_k) \\
&= \sum_{k=1}^{n} \sum m = 1^{\mathrm{n}} \det J_{km} \\
&= n \det J(f).
\end{aligned}
$$

Lemma (ix): *If f is twice continuously differentiable with* $||f(x)|| = 1$, *then* $\nabla \cdot G(f) = 0$.

Proof: This follows from the previous lemma by noting that for such an f and for $0 \le i \le n$, we have $\partial_i |f|^2 = 2f \cdot \partial_i f = 0$. Thus, each of the n columns of the Jacobian of f are orthogonal to f and so are linearly dependent. Hence, $\det J(f) = 0$.

We can now deduce *property (vi)* by applying Green's theorem to $\delta(f)$.

$$\delta(f) = \int_{S^{n-1}} S^{n-1} G(f) \cdot \mathbf{n}\, dS = \int_{B^n} \nabla \cdot G(f)\, dV = 0$$

Let $f, g \in Z$ and suppose $f(x) + g(x) \neq 0$ on S^{n-1}. Fix $\varepsilon > 0$ such that f and g are defined for $1-\varepsilon < |x| < 1+\varepsilon$. Set $1-\varepsilon < \alpha < \beta < 1$. Let φ: $\boldsymbol{R} \to [0, 1]$ be a twice continuously differentiable function defined on $[1-\varepsilon, 1+\varepsilon]$ such that $\varphi(x) = 0$ on $[1-\varepsilon, \alpha]$ and $\varphi(x) = 1$ on $[\beta, 1+\varepsilon]$. Now, we take the convex combination $(1 - \varphi(|x|)f(x) + \varphi(|x|)g(x)$ of f and g and call its normalization $h(x)$. Then, h must also be in the set Z. Thus, $\nabla \cdot G(h) = 0$. Fix $\gamma \in (1-\varepsilon, \alpha)$.

$$\int_{\gamma<|x|<1} \nabla \cdot G(h)\, dV = \int_{\gamma<|x|<1} \nabla \cdot G(f)\, dV$$

From which we obtain, after applying Green's theorem,

$$\int_{|x|=1} G(h) \cdot \mathbf{n}\, dS - \int_{|x|=\gamma} G(h) \cdot \mathbf{n}\, dS$$

$$= \int_{|x|=1} G(f) \cdot \mathbf{n}\, dS - \int_{|x|=\gamma} G(f) \cdot \mathbf{n}\, dS.$$

But, $h = f$ on $|x| = \gamma$ and $h = g$ on $|x| = 1$. From the definition of δ, we obtain the result. That is, $\delta(f) = \delta(g)$.

To reach our goal of using δ to prove Brouwer's theorem we shall need to extend our definition to an arbitrary function f that is continuous and nonzero on $\boldsymbol{S}^{n-1}$.

Definition : *Let f be continuous and nonzero on $\boldsymbol{S}^{n-1}$. Let $g \in Z$ be such that $f(x) \cdot g(x) > 0$ for all $x \in \boldsymbol{S}^{n-1}$. Then, we define $\delta(f) = \delta(g)$.*

It is clear that δ is well defined. If g_1 and g_2 are two functions in Z whose inner products with f are positive on the unit sphere, then so too is $(g_1(x) + g_2(x)) \cdot f(x) > 0$, and so $g_1(x) + g_2(x) \neq 0$. By *property (vii)*, $\delta(g_1) = \delta(g_2)$. It should also be evident that such a function g actually exists. Given f, we can extend f to a neighbourhood of $\boldsymbol{S}^{n-1}$ by setting

$$f(x) = |x| \left(\frac{x}{|x|} \right)$$

Then, f can be approximated by twice continuously differentiable functions using the Weierstrauss Approximation Theorem. Normalizing the resultant function yields a member of Z with the desired property.

With the function δ defined, deducing Brouwer's theorem is fairly straightforward. Let $f : \boldsymbol{B}^n \to \boldsymbol{B}^n$ be continuous. Define $h(x) = x - f(x)$.

Now, let $x \in \boldsymbol{S}^{n-1}$ and suppose $(1-t)x + th(x) = 0$ for some $t \in (0, 1)$. Then, taking the inner product with x tells us

$$(1 - t) + t(x \cdot h(x)) = 0.$$

That is,

$$h(x) \cdot x = \frac{t-1}{t}, \text{ for some } t \in (0, 1].$$

But, $x \cdot h(x) = x(x - f(x)) = 1 - x \cdot f(x) > 0$. This contradiction tells us that x and $h(x)$ nowhere point in opposite directions on the unit sphere. Hence, by *property (vii)*, $\delta[h] = \delta[id] \neq 0$. Then, by *property (vi)*, any continuous extension of h to the whole of the unit ball must vanish somewhere. That is, f must have a fixed point.

2.3 Sperner's lemma and Brouwer's fixed point theorem

We state Brouwer's Fixed Point Theorem again as follows:

***Theorem* [*Brouwer's Fixed Point*]:** *Every continuous function $f: D_n \to D_n$ has a fixed point where $D_n = \{x \in \boldsymbol{R} : |x| \leq 1\}$.*

The purpose of this section is to provide a proof of Brouwer's Fixed Point Theorem for $n = 2$ using a combinatorial result know as Sperner's Lemma and to explore both Sperner's Lemma and Brouwer's Fixed Point Theorem. We begin by examining the case where $n = 1$.

Proof for n = 1: We have $f: [-1, 1] \to [-1, 1]$. Define $g(x) = x - f(x)$. Then $g(1) \geq 0$ and $g(-1) \leq 0$. So, by the Intermediate Value Theorem, there must be a point x_0 so that $g(x_0) = 0$. We, then, have $f(x_0) = x_0$.

Definition: *A fixed point domain is a set $G \subset \boldsymbol{R}^n$ such that every continuous function from G to itself has a fixed point.*

So, Brouwer's Fixed Point Theorem asserts that each D_n, $n \geq 1$, is a fixed point domain.

On the other hand, D_n less a point, [0, 1], [2, 3] and $\boldsymbol{R}$ are not fixed point domains. (It is easy to construct appropriate functions with no fixed point for these sets.) The fact that D_n is compact is important in Brouwer's Fixed Point Theorem.

However, if G is a fixed point domain, and $f : G \to H$ is a continuous bijection with a continuous inverse, then H is a fixed point domain.

Proof: If $g : H \to H$ is continuous, so is $f^{-1} \circ g \circ f$. As this is a continuous function from G to G, it has a fixed point x_0. Now, $(f^{-1} \circ g \circ f)(x_0) = x_0$, so $g(f(x_0)) = f(x_0)$, which means $f(x_0)$ is a fixed point of g.

This means that being a fixed point domain is a topological property. Specifically, Brouwer's Fixed Point Theorem for $n = 2$ is now equivalent to showing that every continuous function from a triangle to itself has a fixed point because there is a continuous bijection with a continuous inverse from the circle to a triangle.

2.3.1 Sperner's Lemma

A triangulation of a triangle is a subdivision of the triangle into small triangles. The small triangles are called *baby triangles* and the corners of the triangles are called *vertices*. The additional condition that each edge between two vertices is part of at most two triangles is necessary.

A *Sperner Labeling* is a labeling of a triangulation of a tri-angle with the numbers 1,2 and 3 such that

(i) The three corners are labeled 1, 2 and 3.
(ii) Every vertex on the line connected vertex i and vertex j is labeled i or
j.

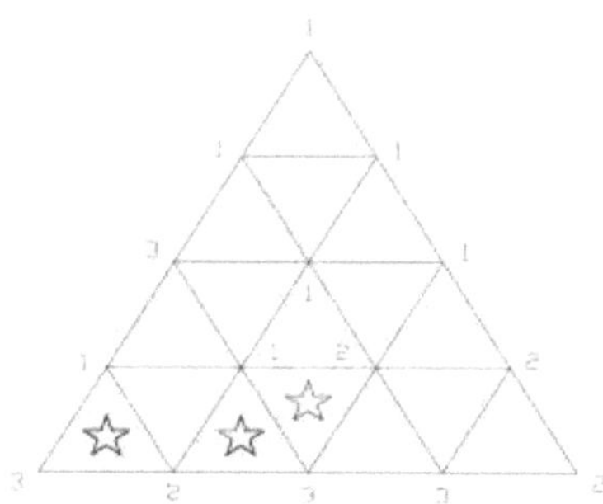

***Lemma*[*Sperner's Lemma*]:** *Every Sperner Labeling contains a baby triangle labeled 1-2-3.*

Sperner's Lemma generalizes to *n* dimensions, where it says that every coloring triangulation of a hypertetrahedron into n + 1 colours must have an baby hyper tetrahedron that is labeled with all n + 1 labels. The proof that follows in fact generalizes to n dimensions too. For details, see *page 34.*

Proof: First, we need the simple fact that the number of 1-2 edges on the outside of the triangle is odd. To see this, label each edge segment on the boundary line between corners 1 and 2 by the difference of the Sperner labels at each end. Then, 1-2 edges are labeled ±1 while the 1-1 and 2-2 edges are labeled 0. Observe that the sum of the labels equals the difference between the corners which is 1. Thus the number of edges labeled ±1 is odd.

We, now, view the triangle as a house. Each baby triangle is a room and each 1-2 edge is a door. We consider a path through the triangle starting from outside the triangle. Note that such a path can only end by leaving the triangle again or by entering a 1-2-3 triangle. Note also that once the entrance is chosen, the path is uniquely determined because no triangle can have three 1-2 edges. This also means that no two paths can meet in a single triangle.

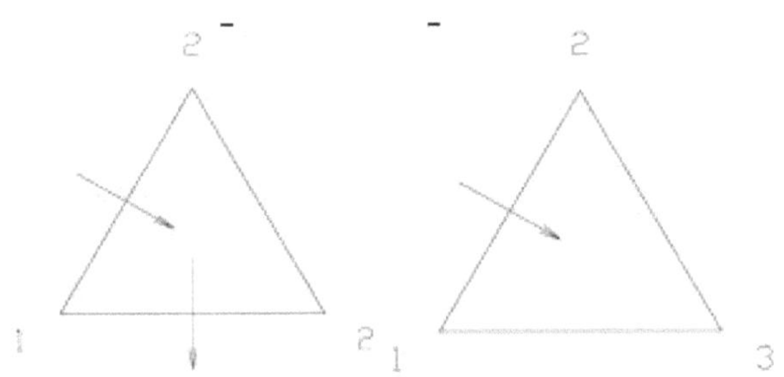

Consider all such paths. A path which ends by exiting the triangle determines a pair of 1-2 edges. Such paths account for an even number of 1-2 edges on the boundary. Since the number of 1-2 edges is odd, there must be at least one 1-2 edge whose path terminates in a 1-2-3 triangle.

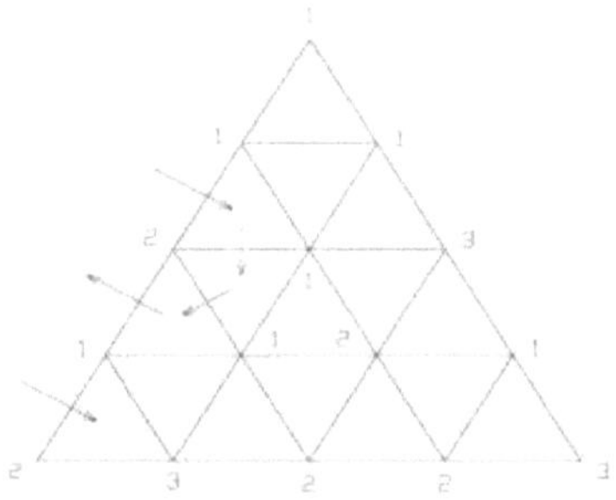

***Lemma* [*Stronger form of Sperner*]:** Without loss of generality, suppose that the outside triangle in a Sperner labeling is labeled 1-2-3 in clockwise order. Let A be the number of baby 1-2-3 triangles oriented in the clockwise direction, and let B be the number of such triangles oriented in the counter clockwise direction.Then, $A-B=1$.

Note that if we prove this, we have that $A+B$ which is total number of baby 1-2-3 triangles, is odd, and thus, greater than 0.

The last proof of Sperner's Lemma can be modified to show this stronger result. The interested reader should try to make this modification.

This result is included in addition to the last proof of Sperner's Lemma because the proofs are beautiful, insightful and significantly different. This next proof is from *page 33.*

Proof: We will label each edge in the triangulation. If an edge has a triangle on either side, it will get two labels, one for each baby triangle it is part of. If we have a baby triangle and two of its vertices (in clockwise direction) are labeled i and j, the label of the edge is 0 if $i = j$ mod 3, 1 if $j = i + 1$ mod 3 and -1 if $j = i - 1$ mod 3.

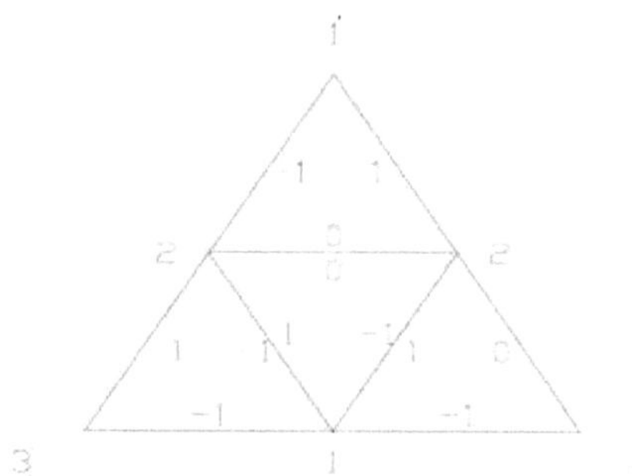

We make three important observations.

(i) The inside edge labeling of a baby triangle sum to 3 if the triangle is 1-2-3 clockwise, if it is 1-2-3 counter-clockwise, and 0 otherwise.
(ii) If an edge participates in two triangles, then its label in one will be the negative of its label in the other.
(iii) If we consider the sum of the inside edge labels of a polygon composed of baby triangles, we get the sums of the inside edge labels of the baby triangles.

The sum of the edges of the big triangle is 3 because each side contributes 1 in total. Since this is also equal to the sum of the sums

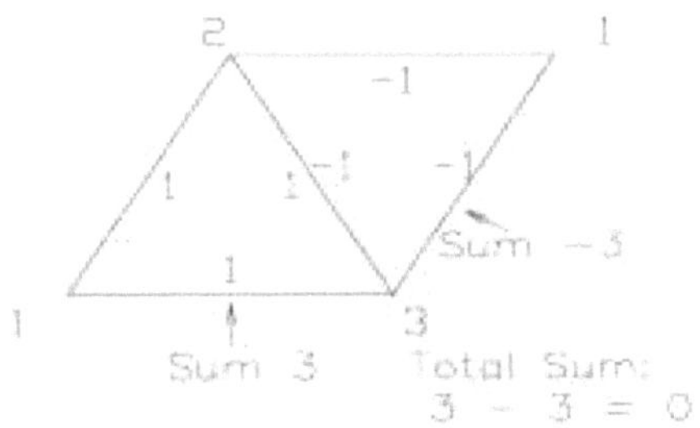

of the inside edge labels of all the baby triangles, we get

$$3A - 3B = 3$$

2.3.2 Barycentric Coordinates

Barycentric coordinates express a point in a triangle as a weighted average of the three vertices. Barycentric coordi- nates are thus always ordered triples (each weight is one number) that add up to 1.

For example, if the triangle is (0, 0), (0, 1), (1, 0), the Barycentric coordinates of the three vertices are (1, 0, 0), (0, 1, 0) and (0, 0, 1) respectively. In this example, the Barycen- tric coordinates of the point(1/4 , 1/4) would be (1/2 , 1/4 , 1/4).

Barycentric coordinates always exist, because a triangle can be thought of as the convex hull its vertices.

2.3.3 Brouwer's Fixed Point Theorem

***Theorem* [*Brouwer's Fixed Point Theorem for n = 2*]:** *Every continuous function from a triangle to itself has a fixed point.*

Although, we won't prove Brouwer for $n > 2$, it is worth noting that both Sperner's Lemma and techniques of this proof generalize to higher dimensions.The proof that follows is from *page 31*.

Proof: Let f be a continuous function from a triangle T into itself. We will write $(a, b, c) \mapsto (a', b', c')$ is $f(a, b, c) = (a', b', c')$ in Barycentric coordinates.

We will label each point in the triangle as follows. Suppose (a, b, c) $\mapsto (a', b', c')$

- If $a' < a$, we label (a, b, c) 1.
- If $a' \geq a$, but $b' < b$ we label (a, b, c) 2.
- If $a' \leq a$ and $b' \geq b$, but $c' < c$ and we label (a, b, c) 3.

If we cannot label the point (a, b, c), then, $a' \geq a$, $b' \geq b$ and $c' \geq c$ so $a = a'$, $b = b'$, $c = c'$ and the point is a fixed point. For the remainder of the proof, we assume that any point we need can be labeled; if not, it is fixed, and we are done.

What we are intuitively doing with this labeling is picking a corner the point does not move closer to. We will eventually get a sequence of small 1-2-3 triangles converging to a point. If this point was not a fixed point, the continuity of f would suggest that as our triangles approach the point, all the corners would have the same labeling. This will be formalized as follows. First, we need to show that this labeling is in fact a Sperner labeling. If we look at the corners, we note

- If $(1, 0, 0) \mapsto (a, b, c)$, $a < 1$ unless this is a fixed point, so this corner gets label 1.

• If $(0, 1, 0) \mapsto (a, b, c)$ then $a \geq 0$ but $b < 1$ (unless this is a fixed point), so this corner gets label 2.
• Similarly, (0, 0, 1) gets label 3.

If we look at a point (a, b, 0) on the line connecting (1, 0, 0) and (0, 1, 0), we see that if $(a, b, 0) \mapsto (c, d, e)$ then $c < a$ or $b < d$, so the label is 1 or 2. If not, we have $c \geq a$, $b \geq d$ so $c = a$, $d = b$ and $e = 0$ and the point is a fixed point.

Similarly, we can show that the points on the line connecting (0, 1, 0) and (0, 0, 1) get labels 2 or 3 and the points on the line connecting (0, 0, 1) and (1, 0, 0) get labels 3 or 1. Thus, if we label a triangulation in this manner we get a Sperner labeling.

We now consider a sequence of triangulations with diame-ter going to 0. (The diameter of a triangulation is defined to be the maximum distance between adjacent vertices in a triangulation.) Each of these triangulations have at least one baby 1-2-3 triangle. Suppose these triangles have vertices

$$(x_{n,1}, y_{n,1}, z_{n,1})$$
$$(x_{n,2}, y_{n,2}, z_{n,2})$$
$$(x_{n,3}, y_{n,3}, z_{n,3})$$

with labels 1, 2 and 3 respectively. Here, the n indicates that the triangle is from the nth traingulation in the sequence of triangulations with diameter going to 0.

We, now, use the Bolzano-Weierstrass Theorem so find a convergent subsequence

$$(x_{n_k,i}, y_{n_k,i}, z_{n_k,i}) \to (x, y, z)$$

for $1 \leq i \leq 3$. We can do this for all three sequences at once because they are the corners of triangles whose diameters tend to 0. If we wanted to avoid the Bolzano-Weierstrass Theorem, an alternate approach is to use sub-triangulations and find nested 1-2-3 triangles.

Now, if

$$\left(x_{n_k,i}, y_{n_k,i}, z_{n_k,i}\right) \mapsto \left(x'_{n_k,i}, y'_{n_k,i}, z'_{n_k,i}\right)$$

And

$$(x, y, z) \mapsto (x', y', z')$$

we have (by the Sperner labeling)

$$x'_{n_k,1} \leq x_{n_k,1}$$

so by continuity

$$x' \leq x$$

Similarly, $y' \leq y$ and $z' \leq z$, so (x, y, z) is a fixed point.

We have now proved a topological result using a combinatorial lemma about triangles. However, the student of algebraic topology may not be so surprised because the topological proof of Brouwer's Fixed Point Theorem and indeed a large part of algebraic topology relies on atriangulating spaces. This result seems to be fundamentally connected with triangles.

2.4 Generalization

In the past century Brouwer's theorem has seen many generalizations. The literature is vast; in this section we outline some of the major results. The first extensions were to specific function spaces. In 1922, Birkhoff and Kellogg showed that compact convex subsets of $C^n[0,1]$ and $L_2[0,1]$ have the fixed point property. Cacciopoli showed the same is true in $C[0, 1]$. Schauder made the leap to Banach spaces and Tychonoff extended this further to locally convex topological vector spaces. The goal of these earlier efforts was to generalize the space $\boldsymbol{E}$ in which the theorem is set. Following this, results were proven for varying conditions on all three components of the fixed point theorem: the space $\boldsymbol{E}$, the set X and the function $f: X \to X$. Many results were obtained that relaxed requirements on one or two of these components while asking more of the others. Some theorems based on certain boundary conditions do not require that f be a self–map. Throughout most of the development, containment of the image set $f(X)$ in a compact set is necessary. However, in the 50's and 60's, it was shown that we merely require the image to be somehow closer to compact than X itself. At some points we pause from our discussion of what types of sets have the topological fixed point property to consider some results on classes of functions for which closed, bounded, convex sets that the fixed point property. The generalizations we present here will be given in the following template as to facilitate comparison.

Theorem : *Let X be a subset of a space $\mathbf{E}$ and $f: X \rightarrow \mathbf{E}$.*

1. *set:*
2. *space:*
3. *function:*

Then [*conclusion*].

2.4.1 Extension to infinite dimension

Having proved Brouwer's theorem, the fact that a nonempty compact convex set in Banach Space has the topological fixed point property is easy to acquire. Whereas, Brouwer's theorem has been derived through a variety of approaches, the proofs of the infinite dimensional version are generally variations on a theme. We start with a continuous function on a compact convex set and approximate this set with one that can be embedded in $\boldsymbol{R}^n$. Then, Brouwer's theorem is applied to get a fixed point x_n. Letting $n \rightarrow \infty$ gives us a sequence $\{x_n\}$ in our original space whose limit point turns out to be the desired fixed point. The different proofs in the literature vary in the method by which the approximation to finite dimensions is done.

2.4.1. (a) Schauder's Fixed Point Theorem

***Theorem* [Schauder 1927] :** *Let X be a subset of* $\boldsymbol{E}$ *and* $f : X \to \boldsymbol{E}$.

1. *X : compact, convex.*
2. $\boldsymbol{E}$ *: Banach space.*
3. *f : continuous self–map.*

Then, f has a fixed point in X.

Proof: Suppose $f : X \to X$ is a continuous function and fix $\varepsilon > 0$. We will construct a function $g : X \to X$ that almost fixes every point of X. To start, by compactness of X, there exists a finite ε–net, $\{a_1, \ldots , a_n\}$ in $\boldsymbol{E}$ that covers X. For each $i \in \{1, \ldots , n\}$, we define the following function.

$$m_i(x) = \begin{cases} \varepsilon - | x - a_i | & , \quad | x - a_i | \le \varepsilon \\ 0 & , \quad | x - a_i | \ge \varepsilon \end{cases}$$

Note that for each $x \in X$, there exists an i such that $m_i(x)$ is non zero. Thus, we can define $g : X \to X$ by

$$g(x) = \frac{\sum_{i=1}^{n} m_i(x) a_i}{\sum_{i=1}^{n} m_i(x)}$$

For any $x \in X$, we have

$$|g(x) - x| = \left| \frac{\sum_{i=1}^{n} m_i(x) a_i}{\sum_{i=1}^{n} m_i(x)} - x \right|$$

$$= \left| \frac{\sum_{i=1}^{n} m_i(x)(a_i - x)}{\sum_{i=1}^{n} m_i(x)} \right|$$

$$\le \varepsilon$$

Let $X_n = \overline{conv}\{a_1, \dots, a_n\}$ and consider $(g \circ f) : X_n \to X_n$. As it is the convex hull of a finite set of points, X_n can be embedded in $\boldsymbol{R}^n$. Thus, by Brouwer's theorem, there is a point x_n that is fixed by $g \circ f$.

$$\begin{aligned} |x_n - f(x_n)| &\leq |x_n - g \circ f(x_n)| + |g \circ f(x_n) - f(x_n)| \\ &= |g \circ f(x_n) - f(x_n)| \\ &\leq \varepsilon \end{aligned}$$

The sequence $\{x_n\}$ is contained in X, a compact set. Thus, we have a convergent subsequence $\{x_{n_k}\}$ with limit x. By completeness, x is in X. Continuity of f tells us that $f\,(x_{n_k})$ converges to $f(x)$. Now, the above inequality tells us $f(x) = x$.

The following, due to Kakutani, illustrates the necessity of compactness in Schauder's theorem. We cannot merely require X to be closed and bounded.

Example A **:** Let $\ell_2\,(\boldsymbol{Z})$ be the space of functions on $\boldsymbol{Z}$ with norm given by $|x| = \max_Z(|x_n|)$ where x_n is the nth coordinate of x.

Consider the unit ball in $\ell_2(\boldsymbol{Z})$. If we write b^n to be the element in this space that is 1 in the nth coordinate and zero everywhere else, for any $x \in \ \ell_2\,(\boldsymbol{Z})$ we can write

$$x = \sum_{n \in Z} x_n b^n.$$

Define $R : \ell_2\,(\boldsymbol{Z}) \to \ell_2\,(\boldsymbol{Z})$ by $(R(x))_n = x_{n-1}$ for each integer n. Then, R is the right-shift operator and is clearly continuous. Next, define $f : \boldsymbol{B} \to \boldsymbol{B}$ by $f(x) = (1 - |x|)b^0 + R(x)$.

Continuity of f follows from the continuity of the norm, and the continuity of R. To see that f is indeed a self–map of the unit ball, let $x \in \boldsymbol{B}$. Then, we have

$$|f(x)| = |(1 - |x|)b^0 + R(x)|$$

$$\leq |(1 - |x|)b^0| + |R(x)|$$
$$= |1 - |x|| + |x|$$
$$= 1.$$

Now, suppose f has a fixed point. Then, there is an x such that $x - R(x) = (1-|x|)b^0$. But, the left hand side can be a multiple of b^0 only when $x = 0$. In this case, we get $0 = x - R(x) = b^0$, a contradiction. Thus, f is a fixed point free mapping of the unit ball in $\ell_2(\boldsymbol{Z})$.

The above result was actually Schauder's first extension of Brouwer's theorem to infinite dimensional space. He also proved the theorem for X closed and convex but with f a compact mapping. Clearly, a continuous mapping on a compact set is compact but if $f : X \to X$ is compact, we need not have compactness of X. For instance, the retraction of the open unit ball in $\boldsymbol{R}^2$ onto the origin is a compact map. This result is sometimes called "*Schauder's second theorem*".

2.4.1 (b) Tychonoff's Fixed Point Theorem

In 1935, Tychonoff proved the following extension of Brouwer's theorem.

Theorem 3.1.2 : *Let X be a subset of a space* **E** *and $f : X \to$* **E**.

1. *X : compact, convex.*
2. **E** *: locally convex topological vector space.*
3. *f : continuous self – map.*

Then, f has a fixed point in X.

Recall that given two open covers $U = \{U_\alpha\}_{\alpha \in A}$, $V = \{V_\beta\}_{\beta \in B}$ of a space X, the second is a refinement of the first if each member V_β is contained in some U_α. Also, U is called *nbd-finite* provided for each $x \in X$, there is a neighbourhood of x that intersects only a finite number of the U_α. A Hausdorff space X is called *paracompact* when every open cover of X admits a nbd-finite refinement. Given

$Y \subset X$, the star of Y with respect to the cover U is the set $St(Y,U) = \cup \ \{U_\alpha : Y \cap U_\alpha \neq \phi\}$. A covering V is called a *star–refinement* of U provided the cover $\{St(V_\beta, V)\}_{\beta \in B}$ is a refinement of U. We remark that every open cover of a paracompact space admits an open star –refinement.

Lemma: *Let* $\mathbf{E}$ *be a Hausdorff topological space,* $X \subset \mathbf{E}$, *and* $f : X \to X$ *continuous. Then,* f *has a fixed point if and only if for each open cover* $\{U_\alpha\}_{\alpha \in A}$ *of* X, *there exists* $x \in X$ *and* U_α *such that both* x *and* $f(x)$ *lie in* U_α.

Proof: The "only if" part if the implication is certainly clear. Next, assume that for every open cover $\{U_\alpha\}$ of X there is an $x \in X$ and α such that both x and $f(X)$ lie in U_α. Suppose $f : X \to X$ is continuous with no fixed points. Since $\boldsymbol{E}$ is Hausdorff, we can find open neighbourhoods W_x and $Vf(x)$ that separate x and $f(x)$. Then, the open cover $\{W_x\}_{x \in X}$ violates our assumption.

With the lemma and the language above, we can now prove ***Tychonoff's theorem***.

Proof: Let $\{U_k\}$ be an open cover of X. Without loss of generality, assume each U_α is convex. We will approximate X with a finite dimensional set X_n and f with a simplicial approximation F and from the fixed point of F, we obtain an x such that $f(x)$ and x lie in the same U_α for some α.

To this end, let $\{W_i\}$ be a star–refinement of $\{U_\alpha\}$ and $W_1, \dots, W_n$ a finite subcover of this refinement. For each k, $0 \leq k \leq n$, select and element from W_k and label it x_k. Now, define $X_k = \overline{conv}\{x_1, \dots, x_k\}$ and set $m = \dim X_k \leq n$. Subdivide W_k into a series of m–simplexes intersecting only in their faces, small enough that each m–simplex is completely contained in $f\ -1(W_k)$ for some k. Let $\{v_0, \dots, v_s\}$ be the combined vertex sets of our simplexes. For each v_i, $f(v_i)$ is contained in one or more W_k. Pick one of these k and define $F(v_i) = x_k$. Next, extend F linearly to all $x \in X_n$. Then, F is a continuous self-map on X_n and thus, by Brouwer's theorem, has a fixed point, say $x \in X_n$.

To complete our proof we show the existence of an α such that both x and $f(x)$ lie in U_α. Indeed, let $x \in f^{-1}(W_k)$. Then, $f(x) \in W_k \subseteq U_\alpha$, where U_α contains $St(W_k, W)$. Now, x lies in some m–simplex. So let $x = \sum \lambda_i y_i$. Then, $F(x) = \sum \lambda_i F(y_i)$, where each $F(y_i)$ gets mapped to an x_j, where $f(y_i) \in W_j$. Since $f(y_i) \in W_k$, this W_j must be a neighbourhood of W_k. That is, $f(y_i) \in \text{st}(W_k, W) \subseteq U_\alpha$. Thus, by convexity, $F(x) \in U_\alpha$. Since x was a fixed point of F, we are done.

Eventually Schauder's second theorem and Tychonoff's theorem were unified to produce the so–called *"Schauder–Tychonoff" fixed point theorem*. The following was proven by Hukuhara in 1950.

Theorem : *Let X be a subset of a space* **E** *and* $f : X \to \mathbf{E}$.

1. *X : convex.*
2. **E** *: locally convex topological vector space.*
3. *f : compact, continuous self–map.*

Then, f has a fixed point in X.

2.4.2 Fixed Point Properties for Closed Bounded Convex Sets

As noted, compactness is important in establishing the topological fixed point theorem in infinite dimensions. In infinite dimensions, the property fails for closed bounded sets. In this section, we pause from our consideration of continuous functions to consider some classes of functions for which closed bounded convex sets have fixed points.

2.4.2 (a) Theorems with Boundary Conditions

The first of the theorems in this genre was by KKM in 1929. They proved that a continuous function on the unit ball in $\boldsymbol{R}^n$ that mapped the unit sphere into the unit ball must have a fixed point. The following theorem is due to Rothe.

Theorem : *Let X be a subset of a space* **E** *and* $f : X \to \mathbf{E}$.

1. *X : closed unit ball.*
2. ***E*** *: Banach space.*
3. *f : compact, continuous and such that $f(\partial X) \subset X$.*

Then, f has a fixed point in X.

Proof: Suppose $f : \boldsymbol{B} \to \boldsymbol{E}$ is a compact continuous function such that $f(\boldsymbol{S}) \subset \boldsymbol{B}$. Let $r : \boldsymbol{E} \to \boldsymbol{B}$ be the projection of $\boldsymbol{E}$ onto $\boldsymbol{B}$. Since the continuous image of a compact set is again compact, we have that $(r \circ f) : \boldsymbol{B} \to \boldsymbol{B}$ is a compact continuous map. Thus, by Schauder's theorem, there is an $x \in \boldsymbol{B}$ such that $r(f(x)) = x$. If $x \in \text{int}(\boldsymbol{B})$, then $f(x) = x$ by the definition of r. Else, x is on the boundary of the ball, and by supposition, $f(x) \in \boldsymbol{B}$. Thus, $r(f(x)) = f(x)$ and again, x is the desired fixed point of f.

We note that Rothe's theorem does not hold if we simply require that f be a compact mapping on the boundary of X. The following example illustrates this.

Example B **:** Let X be the ball of radius 2 centered at the origin in the space 2. Define $f : X \to X$ as follows. For $x = (x_1, \ldots , x_n, \ldots)$,

$$f(x) = \begin{cases} \left(\sqrt{1 - \|x\|^2}, x_1, x_2, \ldots\ldots, x_n, \ldots\ldots\right), & \|x\| \le 1 \\ (2 - \|x\|) f\left(\dfrac{x}{\|x\|}\right), & 1 \le \|x\| \le 2 \end{cases}$$

We see that f maps the boundary of X to the origin and thus, is a compact mapping on this set. However, a similar argument to that in *example A* shows that f has no fixed points.

Since Rothe proved this result, many authors have investigated various boundary conditions. It is possible to replace the condition $f(\partial X) \subset X$ with any of:

i) $\langle f(x), x \rangle \le \|x\|^2$ (Krasnoselskii),
ii) $\|x - f(x)\| \ge \|f(x)\|$ (Petryshyn) or

iii) $||x - f(x)||^2 \geq ||f(x)||^2 - ||x||^2$ (Altman), for x on the unit sphere.

2.4.2 (b) Conditions on Compactness

Another avenue for the generalization of Brouwer's theorem is through the requirements on compactness. It has already been shown that we do not need the set X itself to be compact as long as the image of X is contained in a compact subset of $\boldsymbol{E}$. It is interesting that what we need is that f in some sense takes X a little closer to being compact. We will consider two types of mappings that satisfy this condition.

Definition: *Let X be a complete metric spac, and M the family of bounded nonempty subsets of X. The Kuratowski measure of noncompactness is the function* $\alpha : M \rightarrow \boldsymbol{R}^+$ *defined by*

$$\alpha(A) = \inf\{\varepsilon > 0 : A \text{ can be covered by a finite number of sets in } M \text{ with diameter } < \varepsilon\}$$

Example C: In infinite dimensional Banach space, $\alpha(\boldsymbol{B}) = 2$.

Clearly, $\alpha(\boldsymbol{B}) \leq 2$. Let $\{e^i\}_{i \in I}$ be a basis for the space where $e^i_j = 1$ if $j = i$, and zero otherwise. Each e^i lies in $\boldsymbol{B}$. The smallest ballcontaining n basis vectors $e^{i_1}, \ldots, e^{i_n}$ would be centered at $\frac{1}{n}\sum_{t=1}^{n} e^{i_t}$ with radius $\sqrt{1 - \frac{1}{n}}$. Then, each ball of diameter $2\sqrt{1 - \frac{1}{n}}$ can cover at most n basis vectors. Thus, it would take an infinite number of these balls to cover $\boldsymbol{B}$. Since any set of diameter $2\sqrt{1 - \frac{1}{n}}$ is contained in a ball of diameter $2\sqrt{1 - \frac{1}{n}}$, any covering of $\boldsymbol{B}$ of sets with diameter $2\sqrt{1 - \frac{1}{n}}$ would have to be infinite. Thus, $\alpha(\boldsymbol{B}) > 2$ $\sqrt{1 - \frac{1}{n}}$ for all n. Thus, $\alpha(\boldsymbol{B}) = 2$.

There are other measures of non compactness. For example, in the above definition we could have taken the infimum over all balls of radius less than ε rather than sets of diameter less than ε. Then, we would have obtained the Hausdorff measure of non compactness. Any measure would do in what follows. We have the following properties (without proof) :

Properties of the Measure of Noncompactness

1. $\alpha(A) = 0 \Leftrightarrow \overline{A}$ is compact.
2. $A \subseteq B \Rightarrow \alpha(A) \leq \alpha(B)$
3. $\alpha(A) = \alpha(\overline{A})$
4. $\alpha(A \cup B) \leq \max\{\alpha(A), \alpha(B)\}$
5. $\alpha(A + B) \leq \alpha(A) + \alpha(B)$
6. $\alpha(cA) = |c|\alpha(A)$
7. $\alpha(\text{conv}(A)) = \alpha(A)$

Definition: *Let X be a complete metric space. A continuous function $f : X \to X$ is called an α–set contraction provided there exists a $k \in [0, 1)$ such that for all $A \subseteq X$, we have $\alpha(f(A)) \leq k\alpha(A)$.*

Clearly, any contraction mapping is an α–set contraction. Furher, any completely continuous com pact function is an α–set contraction.This follows since if A is a bounded sub-set of $\boldsymbol{E}$, then $f(A)$ is precompact. Thus, $\alpha(f(A)) = 0 \leq k\alpha(A)$ for any $k \in [0, 1)$.

We can now prove the following theorem by Darbo .

***Theorem* [*Darbo 1955*] :** *Let X be a subset of a space $\boldsymbol{E}$ and $f : X \to \boldsymbol{E}$.*

1. *X : closed, bounded, convex.*
2. *$\boldsymbol{E}$: Banach space.*
3. *f : an α–set contraction self – map on X.*

Then, f has a fixed point in X.

Proof: Let $X_0 = X$ and define $X_n = \overline{conv}(f(X_{n-1}))$ for $n \geq 1$. By convexity of X, $X_1 \subseteq X_0$. An induction argument shows that $X_n \subseteq X_{n-1}$ for each n. Thus, $\{X_n\}_{n=0}^{\infty}$ is a decreasing sequence of closed convex sets. Thus, $X_\infty = \bigcap_{n=0}^{\infty} X_n$ is nonempty, closed and convex. Also,

$$\begin{aligned}\alpha(X_n) &= \alpha(\overline{conv}\ (f(X_{n-1})))\\ &= \alpha(f(X_{n-1}))\\ &\leq k\alpha(X_{n-1})\end{aligned}$$

where k is some fixed real in [0, 1). From this, we obtain $\alpha(X_n) \leq k^n\alpha(X)$. Since X is bounded, $\alpha(X) < \infty$. Taking limits gives $\alpha(X_\infty) = 0$ and so, X_∞ is compact. Further, since $\bigcap_{n=0}^{\infty} X_n = \bigcap_{n=0}^{\infty} \overline{conv}(f(X_n))$, it follows that f is a self-map on X_∞. Thus, by Schauder's theorem f has a fixed point.

Now, we turn to another more general class of functions for which closed, bounded, convex sets have the fixed point property.

Definition **:** *Let X be a complete metric space. A function $f: X \to X$ is called condensing or densifying provided for each noncompact subset A of X, we have $\alpha(f(A)) < \alpha(A)$.*

Sadovski used Zorn's lemma and Schauder's theorem to prove a theorem analogous to but stronger than Darbo's theorem. It is strange that Darbo's theorem is more frequently referenced in the literature. We have included both results for the sake of completeness. To obtain Sadovski's theorem, we need the following result (without proof).

Theorem : *Let X be compact and $f: X \to X$. Then, there exists a nonempty subset $M \subseteq X$ such that $M = \overline{f(M)}$*

Theorem* [*Sadovski 1967*]:** *Let X be a subset of a space* ***E *and $f: X \to$* ***E****.*

1. *X : closed, bounded, convex.*
2. ***E*** *: Banach space.*
3. *f : a condensing self–map.*

Then, f has a fixed point in X.

Proof: Fix $x \in X$. Let $K = \bigcup_{n=1}^{\infty} \{f^n(x)\}$. Since f is condensing, if $\alpha(K) \neq 0$, we have $\alpha(K) = \alpha(f(K) \cup \{f(x)\}) \leq \alpha(f(K)) < \alpha(K)$.

Thus, $\alpha(K) = 0$ and so, $\overline{K}$ is compact. Also, $f(\overline{K}) \subseteq \overline{K}$. Thus, by the above theorem, there is a nonempty subset $M \subset \overline{K}$ such that $M = \overline{f(M)}$. We use M to produce a compact convex set to which we can apply Schauder's theorem. To this end, set

$$C = \{B \subseteq X : M \subseteq B,\ B \text{ closed, convex and invariant under } f\}.$$

Note that C is nonempty since $X \in C$. Next, let $A = \bigcap_{B \in C} B$. Then, $A = \overline{conv}(f(A))$ and assuming $\alpha(A) \neq 0$, we have

$$\alpha(A) = \alpha(\overline{conv}(f(A))) = \alpha(f(A)) < \alpha(A).$$

Thus, $\alpha(A) = 0$ and A is compact. Hence, by Schauder's theorem, f has a fixed point.

2.4.3 Other Extensions

We end this section with a look at fixed point theorems involving more than one function. The geometric counterpart of the topological fixed point theorem is of course the famous Banach contraction principle(will be discussed in the net chapter). The following theorem due to Krasnoselskii considers the fixed points of the sum of a compact continuous mapping with a contraction.

Theorem* [*Krasnoselskii*] :** *Let X be a subset of a space* **E** *and f, g : $X \to$* ***E*.*

1. *X : closed, convex.*
2. ***E*** *: Banach space.*
3. *f and g : continuous, f compact, g a contraction and f(X) + g(X) $\subseteq X$.*

Then, $f + g$ has a fixed point in X.

Lemma: *Let $g : X \to$* ***E*** *be a contraction and id the identity map. Then, $id-g : X \to (id - g)(X)$ is a homeomorphism and if $(id - g)(X)$ is precompact, then so is X.*

***Proof* [*Krasnoselskii*] :** Fix $a \in X$. The the function $x \mapsto g(x) + f(a)$ is a contraction on X and hence, has a unique fixed point x_0 in X. Then, $x_0 - g(x_0) = f(a)$ and so, $(id - g)^{-1} \circ f(a)$ is in X for each a. By the above lemma, $(id - g)^{-1} \circ f$ is continuous and compact and so, has a fixed point z. Then, $f(z) + g(z) = z$.

In the above theorem, the requirement that $f(x) + g(y)$ be in X for all $x, y \in X$ is quite strong. It is natural to ask whether we may merely require $f + g$ be a self map on X. The answer is yes if we assume X is bounded. For, if f is compact, then for $A \subseteq X$, $f(A)$ is precompact. Hence, $\overline{f(A)}$ is compact and $\alpha(f(A)) = 0$. Further, if g is a contraction and $U_1, \ldots, U_n$ is a finite cover of A with $\text{diam}(U_i) < \varepsilon$, then $g(U_1), \ldots, g(U_n)$ covers $g(A)$. For $g(x), g(y) \in g(U_i)$, we have $\|g(x) - g(y)\| \leq k\|x - y\| < k\varepsilon$, for some $k \in [0, 1)$. Thus, $\text{diam}(g(U_i)) < \text{diam}(U_i)$ and so, $\alpha(g(A)) < \alpha(A)$. In summary,

$$\alpha((f + g)(A)) = \alpha(f(A) + g(A)) \leq \alpha(f(A)) + \alpha(g(A)) < \alpha(A).$$

Thus, $f + g : X \to X$ is condensing and by Sadovski's theorem, $f + g$ has a fixed point.

Much theory has been developed on fixed point theorems for families of continuous functions. The following was obtained by

Markov using the Tychonoff fixed point theorem. Kakutani gave a direct proof of the theorem for a family of equicontinuous functions. Recall that a family of functions $F = \{f\}$ is said to be *equicontinuous* on a set X if for every $x \in X$ and any family of neighbourhood $\{V_f\}_{f \in F}$ of $f(x)$, there exists a neighbourhood U of x such that $y \in U$ implies $f(y) \in V_f$, for all $f \in F$.

A family of functions $F = \{f\}$ defined on a set X is said to have a fixed point if there is a mutual fixed point for all $f \in F$. A family F is said to be *commuting* if $f \circ g = g \circ f$ for all $f, g \in F$.

Theorem **[*Markov 1936*] :** *Let X be a subset of a space **E** and $f: X \to$ **E**.*

1. *X : compact, convex.*
2. ***E** : locally convex topological vector space.*
3. *$F = \{f\}$: a commuting family of affine functions.*

Then, F has a fixed point in X.

Proof: For $f \in F$, let Fix(f) denote the set of fixed points of f. Since f is affine, it is not hard to see that Fix(f) is closed and convex and so, Fix(f) inherits compactness from X. By Tychonoff's theorem, Fix(f) is nonempty. Next, note that for $x \in$ Fix(f) and $g \in F$, $g(x) = g(f(x)) = f(g(x))$. So, $g(x) \in$ Fix(f) and g is a self-map of the compact convex set Fix(f). Thus, as above, the set of points in Fix(f) that are fixed by g is compact, convex and nonempty. This set is exactly Fix(g) $\cap$ Fix(f). Then, by induction, the set {Fix(f) : $f \in F$} has the finite intersection property. Now, by compactness of X, we see

$$\bigcap_{f \in F} Fix(f) \neq \phi$$

Hence, F has a fixed point.

2.5 Multifunctions and Kakutani's Theorem

Topological fixed point theory for multifunctions has its seeds in von Neumann's work in the proof of the fundamental theorem of game theory. In an effort to simplify the arguments used, Kakutani proved an analog of Brouwer's theorem. Since its conception, Kakutani's theorem has been applied numerous times in game theory, economic theory, and control theory. In particular, Nash used it to prove the existence of an equilibrium for finite games.

2.5.1 Kakutani's Fixed Point Theorem

Recall that a multifunction $F : X \to 2^X$ is a set valued map on X. As in Brouwer, we will consider sets that are compact and convex. The basic requirement for a multifunction in topological fixed point theorems is that it be upper semicontinuous with nonempty, compact, convex images. For short, we say F is a *cusco*. An element $x \in X$ is a fixed point for F provided $x \in F(x)$. If every *cusco* mapping X into 2^X has a fixed point, then we say X has the Kakutani fixed point property (*kfpp*).

Lemma : *The Kakutani fixed point property is preserved under retractions.*

Proof: Suppose X has the Kakutani fixed point property. If $r : X \to Y$ is a retraction and $F : Y \to 2^Y$ is a *cusco*, then $G : X \to 2^Y$ defined by $G(x) = F(r(x))$ is a *cusco* and has a fixed point that must also be a fixed point of F.

***Theorem* [*Kakutani*]:** *Let X be a compact convex subset in $\mathbf{R}^n$. If $F : X \to 2^X$ is a cusco, then F has a fixed point.*

Proof: In light of the above lemma, we may assume that X is a k-simplex. Let $n > 0$ be given. Subdivide X into a k-simplicial complex such that each k-subsimplex is contained in a ball of radius $1/n$. Define a single valued function $f_n : X \to X$ as follows. For each vertex v in X, select $f_n(v)$ in $F(v)$. Then, extend f_n linearly over each k-subsimplex in the complex. So, f_n is continuous and by Brouwer's theorem, has a fixed point, say x_n. Consider the sequence $\{x_n\}$ in X. We may assume $\lim_{n\to\infty} x_n = x$ since otherwise we can consider a convergent subsequence. We will show that x is our desired point.

To this end, let S_n be a k-simplex in our complex containing x_n and let $x_0^n, \ldots, x_k^n$ be the vertces of S_n. As $n \to \infty$, the diameter of S_n goes to zero. Thus, $x_i^n \to x$ for each i. Let $f_n(x_i^n) = y_i^n$. So, we can write $x_n = \sum_{i=0}^{k} \lambda_i^n y_i^n$, where $\sum_{i=0}^{k} \lambda_i^n = 1$, $\lambda_i^n \geq 0$. We can assume without loss of generality that $x_n \to x$, $\lambda_i^n \to \lambda_i$ and $y_i^n \to y_i$ all converge. Otherwise, we could use convergent subsequentces. Since for each n, $\sum_{i=0}^{k} \lambda_i^n = 1$, $\lambda_i^n \geq 0$, we must have $\sum_{i=0}^{k} \lambda_i = 1$, $\lambda_i \geq 0$. So, $x = \sum_{i=0}^{k} \lambda_i y_i$ is a convex combination. Further, since $x_i^n \to x$, $y_i^n \to y_i$, $y_i^n \in F(x_i^n)$, by upper semicontinuity of F, we must have $y_i \in F(x)$ for each i. Thus, by convexity of $F(x)$, we obtain $x \in F(x)$.

Not surprisingly, the *kfpp* is very closely related to the *tfpp*. If X has the *kfpp*, then since any continuous single valued function can be viewed as a cusco, X will also have the *tfpp*. Further, if X is a compact subset of Euclidean space, then the two properties are equivalent. For, in this case, Cellina's theorem says that there is an approximate continuous selection f of F. That is, for each $\varepsilon > 0$, there exists $f_\varepsilon : X \to X$ continuous with

$$d_{Gr(F)}(x, f_\varepsilon(x)) < \varepsilon ,$$

for any x in X. So, if X has the *tfpp*, then each f_ε has a fixed point, say x_ε. Letting ε go to zero gives a sequence $(x_\varepsilon, x_\varepsilon)$ that approaches the graph of F. Since F is a cusco, this graph is closed. Hence, $\lim_{\varepsilon \to 0} (x_\varepsilon, x_\varepsilon) = (x, x) \in Gr(F)$. That is, x is a fixed point for F. It follows that the two properties are also equivalent for the class of convex sets. For, if X is convex and non compact, then X doesn't have the *tfpp* (or the *kfpp*).

2.5.2 Extended to Banach Space

As in the case for single valued mappings, a set that can be approximated by a sequence of sets having the Kakutani fixed point property will inherit this property. It is through this approximation process that we can extend Kakutani's theorem to Banach space. As stated, Bohnenblust and Karlin first showed the extension of Kakutani's theorem to Banach space.

Lemma : *Let **E** be a normed spac, and $X \subset$ **E** a compact and convex. Suppose that for every $n \in$ **N**, there exists $f_n : X \to X$ linear such that*

1. $||f_n(x) - x|| < \varepsilon_n$ *for all $x \in X$, where $\varepsilon_n \to 0$ as $n \to \infty$.*
2. *$f_n(X)$ has the Kakutani fixed point property for each n.*

Then, X has the Kakutani fixed point property.

Theorem : *The Hilbert cube $H = \{(x_n) \in \ell_2 : |xn| \leq 1/n \}$ has the kfpp.*

Theorem : *Let X be a compact convex subset in Banach space. There is a linear homeomorphism of X onto a compact convex subset of H.*

Theorem : *Let X be a compact convex subset of a Banach space and $F : X \to 2^X$ a cusco. Then, F has a fixed point.*

Theorem : *Let X be a closed convex subset of a Banach space and $F : X \to 2^X$ a cusco. If F(X) is contained in a compact subset of X, then F has a fixed point.*

2.5.3 Theorems with Boundary Conditions

In this section we take a slight detour to consider some results in which the function considered is not strictly a self–map. The first theorem we consider is immediate from the material in the previous section. In the second part, we look at some interesting results by Browder that don't explicitly make any requirements that F map X into 2^X. Of course, it is a necessary consequence of the conditions that Browder does insist on that $F(X) \cap X \neq \phi$. This theorem is of particular interest because it gives better insight into the characteristics of F that give us a fixed point. As well, the preliminary results need to prove the theorems of Browder are of interest in their own right.

We begin with the multifunction analog of Rothe's theorem.

Theorem : *Let X be a closed convex subset of a Banach space $\mathbf{E}$ and $F : X \rightarrow 2^{\mathbf{E}}$ a cusco such that $F(X)$ is precompact. If $F(x) \subseteq X$ for each $x \in \partial X$, then F has a fixed point.*

Proof: First note that if X has empty interior, then $X = \partial X$ and the result follows from the last theorem of section 2.5.2. If $\text{int}(X) \neq \phi$, then we can assume without loss of generality that $0 \in \text{int}(X)$. By virtue of $F(X)$ being contained in a compact subset of X, choose $c \in \boldsymbol{R}^+$ such that $F(X) \subseteq cX$. Next, let g be the Minkowski functional and let $r : cX \rightarrow X$ be the retraction defined by $r(x) = (\max\{g(x), 1\})^{-1}X$. Consider the function $G : cX \rightarrow 2^{cX}$ defined by $G(x) = F(r(x))$.

Then, G is a *cusco* and so there is an x_0 cX with $x_0 \in G(x_0)$. Note that if $x \notin X$, then $r(x) \in \partial X$ and so $G(x) \subseteq X$. Thus, any fixed point of G must be in X. It follows that x_0 is a fixed point of F.

Now we turn to our first result by Browder. The previous multifunction results have required F to be *usc*. In the following, we ask that the inverse image of open sets under F be open. In this case, F is said to be lower semicontinuous.

Theorem : *Let X be a nonempty compact convex subset of a topological vector space $\mathbf{E}$ and $F : X \rightarrow 2^X$ such that $F(x)$ is nonempty and convex for each $x \in X$. Suppose that $F^{-1}(y) = \{x \in X : y \in F(x)\}$ is open for each $y \in X$. Then, F has a fixed point.*

Proof : The set $\{F^{-1}(y) : y \in X\}$ is an open cover of our compact set X. Thus, choose a finite subcover $\{F^{-1}(y_i)\}_{i=1}^{n}$, and let $p_i : X \to \boldsymbol{R}$, $1 \le i \le n$, be a partition of unity subordinate to this cover. Define $p : X \to X$ by

$$p(x) = \sum_{i=1}^{n} p_i(x)y_i.$$

By convexity of X, p is well-defined. Also, for each i such that $p_i(x) \neq 0$, we have $x \in F^{-1}(y_i)$. Thus, $y_i \in F(x)$. Since $\sum_{i=1}^{n} pi(x) = 1$ and $p_i(x) \geq 0$, we see that $p(x)$ is a convex combination of points in $F(x)$. Since $F(x)$ is convex, $p(x) \in F(x)$. But, p is a continuous self-map on the finite dimensional set. So, by Brouwer's theorem, p has a fixed point. This point is the desired fixed point for X.

Lemma 1 : *Let X be a compact subset of a topological vector space $\boldsymbol{E}$ and $f : X \to \boldsymbol{E}^*$ continuous. Then, for a fixed $y \in X$, the function $g : X \to \boldsymbol{R}$ defined by $g(x) = \langle f(x), x-y \rangle$ is continuous.*

Lemma 2 : *Let $\boldsymbol{E}$ be a locally convex topological vector space and X a nonempty subset of $\boldsymbol{E}$. Let $F : X \to 2^{\boldsymbol{E}}$ be a cusco and suppose F has no fixed points. Then, there exists a continuous function $f : X \to \boldsymbol{E}^*$ such that for $x \in X$, $f(x)$ is strictly positive on $x - F(x)$.*

Lemma 3: *Let X be a compact convex subset of a locally convex topological vector space $\boldsymbol{E}$ and $f : X \to \boldsymbol{E}^*$ continuous. Then, there exists $x \in X$ such that $\langle f(x), x - y \rangle \geq 0$* for all $y \in X$.

Theorem : *Let $\boldsymbol{E}$ be a locally convex topological vector space and X a nonempty compact convex subset of $\boldsymbol{E}$. Let $F : X \to 2^{\boldsymbol{E}}$ be a cusco and suppose that for each $x \in \partial X$, there exists $y \in F(x)$, $z \in X$, $\lambda < 0$ such that $y - x = \lambda(z - x)$. Then, F has a fixed point.*

Proof: Suppose the contrary. Then by *Lemma 2*, there is a continuous $f : X \to \boldsymbol{E}^*$ such that $f(x)$ is strictly positive on $x - F(x)$. By *Lemma 3*, there exists $x_0 \in X$ such that

$$\langle f(x_0), x_0 - x\rangle \geq 0, \text{ for all } x \in X. \ldots\ldots\ldots\ldots\ldots (*)$$

There are two cases to consider.

First, suppose $x_0 \notin \partial X$. Since X is convex, for $y \in \boldsymbol{E}$, there is an $\varepsilon > 0$ with $x = x_0 - \varepsilon\ y \in X$. We have $0 \leq \langle f(x_0), x - x_0\rangle = \varepsilon \langle f(x_0), y\rangle$. Thus, $\langle f(x_0), y\rangle \geq 0$ for all $y \in \boldsymbol{E}$. Replacing y with $-y$, we obtain $\langle f(x_0), y\rangle = 0$ for all $y \in \boldsymbol{E}$. On the other hand, (x_0) is strictly positive on the non empty set $x_0 - F(x_0)$ and so, we have a contradiction.

Now suppose $x_0 \in \partial X$. Then, by our assumption, choose $y \in F(x_0)$, $x \in X$, $\lambda < 0$ with $y - x_0 = \lambda(x - x_0)$. Since $f(x_0)$ is strictly positive on $x_0 - F(x_0)$, we have $0 < \langle f(x_0), x_0 - y\rangle = \lambda\langle f(x_0), x_0 - x\rangle \leq 0$, by (*) above, and since $\lambda < 0$. Thus, again we have a contradiction. It follows that F must have a fixed point.

The corresponding result for inward mappings is easily ob- tained from the previous theorem.

Theorem : *Let* $\mathbf{E}$ *be a locally convex topological vector space and* X *a nonempty compact convex subset of* $\mathbf{E}$*. Let* $F : X \to 2^E$ *be a cusco and suppose that for each* $x \in X$*, there exists* $y \in F(x)$*,* $z \in X$*,* $\lambda > 0$ *such that* $y - x = \lambda(z - x)$*. Then,* F *has a fixed point.*

Proof: We reduce this to the previous case of the above theorem. Define $G : X \to 2^E$ by $G(x) = 2x - F(x)$. Then, G is a *cusco*. Now, for $x \in \partial X$, choose $y \in F(X)$, $z \in X$, $\lambda > 0$ such that $y - x = \lambda(z - x)$. Then, $2x - y \in G(x)$ and $(2x - y) - x = x - y = -\lambda(z - x)$ with $-\lambda < 0$. Thus, G satisfies the hypothesis for the previous theorem and so, has a fixed point, say x. Then, $x \in 2x - F(x)$. It follows that x is a fixed point for F.

2.6 Applications

The application of topological fixed point theory is vast. In fields such as economics and game theory it is the only known method by which many important results are obtained. Brouwer can be used to show the existence of a winner in the game of Hex and has been useful in other two player mathematical games. The theory is also fundamental in showing the existence of solutions for differential equations. As well, Brouwer's theorem and its relatives have served to simplify existing proofs of many well known results. In this section, we highlight a few of these applications.

2.6.1 The KKM-Map Principle

In *section 2.2,* we saw that Sperner $\Rightarrow$*KKM* $\Rightarrow$ Brouwer. The ease in which this chain was established caused much speculation as to their equivalence. The question was open for almost 50 years when in 1974 Yoseloff showed that Brouwer $\Rightarrow$ Sperner. This is perhaps one of the most important relationships Brouwer's theorem has with any result. Together, the three theorems form a mathematical trinity that has had a great number of applications. The *KKM* theorem was extended to topological vector spaces in 1952 by Ky Fan. Its extensions and equivalent formulations have had such an effect as to have spawned a branch of research known as *KKM* theory. The theorem itself has become known as the *KKM* – map principle and has been widely used as a tool for fixed point theory, minimax problems, dimension theory and mathematical economics. As such, we consider Brouwer $\Rightarrow$ *KKM* to be an important application of the Brouwer fixed point theorem.

***Theorem* [*KKM*]:** *Let X be a nonempty subset of Euclidean space E. Suppose that for every $x \in X$, there exists a closed subset $M(x) \subseteq X$ such that*

$$conv F \subseteq \bigcup_{x \in F} M(x)$$

for all finite subsets F of X. Then, for any finite subset F of X,

$$\bigcap_{x \in F} M(x) \neq \phi .$$

Hence, if for some $x \in X$, $M(x)$ is compact, we have

$$\bigcap_{x \in X} M(x) \neq \phi .$$

Proof: Suppose the contrary. That is, suppose the hypothesis of the theorem and that $F = \{x_1, \ldots , x_m\}$ is a finite set in X with $\cap_{i=1}^{m} M(x_i) = \phi$.

Consider f : conv(F) $\rightarrow E$ defined by

$$y \mapsto \frac{\sum_{i=1}^{m} d_{M(x_i)}(y) x_i}{\sum_{i=1}^{m} d_{M(x_i)}(y)}$$

This map is well defined since by assumption $\cap_{i=1}^{m} M(x_i) = \phi$. Thus, we have f is a self–map on conv(F). By Brouwer's theorem, f has a fixed point, say z.

Define $F' = \{x \in F : z \notin M(x)\}$. Now,

$$z = \frac{\sum_{i=1}^{m} d_{M(x_i)}(z) x_i}{\sum_{i=1}^{m} d_{M(x_i)}(z)}$$

Note that $z \notin M(x) \Leftrightarrow d_{M(x)}(z) \neq 0$. Therefore, z can be written as a convex combination of elements in conv(F'). From the assumed property, conv(F') $\subseteq \cup_{x \in F'} M(x)$. This implies $z \in M(x)$ for some $x \in F'$, a contradiction. Thus, $\cap_{i=1}^{m} M(x_i) = \phi$.

The final statement of the theorem follows using the finite intersection property characterization of compactness.

Multifunctions satisfying the hypothesis of the KKM theorem, that is, functions $F: X \to 2^X$ with $conv\{x_1, \dots, x_n\} \subseteq \cup_{i=1}^{n} F(x_i)$, for finite subsets $\{x_1, \dots, x_n\}$ of X are called KKM–maps. For example, let $\boldsymbol{E}$ be a normed linear space, X a convex subset and $f: X \to \boldsymbol{E}$ a continuous function. Then, it can be shown that $G: X \to 2^X$ defined by

$$G(x) = \{ y \in X : ||f(y) - y|| \leq ||f(y) - x|| \}$$

is a *KKM*–map. Thus, the *KKM*–map principle is useful in showing the existence of best approximations.

2.6.2 Solutions to Differential Equations

The usefulness of fixed point theorems in proving the existence of solutions for differential equations cannot be overstated. Often, the solution of an equation is found to coexist with the existence of a fixed point of some related function. For example, a solution to the equation $F(x) = 0$ is a fixed point of $F(x) + x$. If the set of valid solutions can be viewed as a compact convex subset of some space, then it may be possible to show the existence of a fixed point using topological fixed point theorems. Often in this area, the space is a function space and so Schauder's theorem may be helpful.

As an example, we show the existence of a continuous $y: \boldsymbol{R} \to \boldsymbol{R}^n$ with

$$y'(t) = f(t, y(t))$$
$$y(a) = b$$

where f is a continuous function of t and y in a neighbourhood of (a, b).

First, note that by continuity, we can assume

$$|f(t, y)| \leq K, \text{ for } |t-a| \leq \varepsilon \text{ and } |y-b| \leq \varepsilon \ .$$

Let $\delta < \min \{ \varepsilon, \frac{\varepsilon}{K} \}$. Consider the space $\boldsymbol{E}$ of functions from $\boldsymbol{R}$ into $\boldsymbol{R}^n$ that are continuous on $|t-a| < \delta$. We equip $\boldsymbol{E}$ with the uniform norm. The set to which we will apply Schauder's theorem is

$$X = \{ y \in \boldsymbol{E} : |y(t) - b| \leq \delta K, \text{ for all } t \}.$$

It is not hard to see that M is closed and convex. Next, define $h : X \rightarrow X$ by

$$h(y)(t) = b + \int_a^t f(s, y(s))ds.$$

Then, for $y \in X$, $|h(y)(t)-b| \leq |t-a| \max \{|f(s, y(s))|\} \leq \delta K$. Thus, h is indeed a self–map of X. Also, h is continuous since

$$|h(y_1) - h(y_2)| = \sup_t \left| \int_a^t f(s, y_1(s)) - f(s, y_2(s)) \right|$$

approaches zero as $|y_1 - y_2| \rightarrow 0$ by uniform continuity of f. Next, note that for any $y \in X$, $|h(y)| \leq |b| + \delta K$ and so, $h(X)$ is uniformly bounded. Also, note that $h(X)$ is equicon- tinuous since

$$|h(y)(t_1) - h(y)(t_2)| \leq \left| f(s, y(s))ds \right| \leq K|t_1 - t_2|.$$

It follows that $h(X)$ is precompact and so, we can apply Schauder's (second) theorem to see that h has a fixed point. This fixed point solves the equation.

2.6.3 The Jordan Curve Theorem

Many authors have employed Brouwer's theorem to give a simplified proof of a known result in mathematics. One such application is Maehara's proof of the Jordan Curve theorem. A Jordan curve is the image under a homeomorphism of a circle. The

Jordan curve theorem says that in $\boldsymbol{R}^2$, the compliment of a Jordan curve J consists of two components, each of which have J as its boundary. This result is notoriously difficult to prove. The first rigorous proof appeared in 1905 . In 1984, Maehara used Brouwer's theorem to give a simplified version of a proof by Moise. Still, a simplified proof of the Jordan curve theorem need not be brief. We present here the lemmas Maehara used to bypass some of the tedious arguments in Moise's proof.

Let J be a Jordan curve in $\boldsymbol{R}^2$. By $E(a, b; c, d)$, we mean the rectangular region

$$\{(x, y) : a \leq x \leq b,\ c \leq y \leq d\}.$$

Theorem : *If* $\boldsymbol{R}^2 - J$ *is not connected, then each of its components has J as its boundary.*

Proof: Let U be a component of $\boldsymbol{R}^2 - J$. Then, for any other component W, we have $\overline{U} \cap U^c \cap W = \phi$. Thus, $\overline{U} \cap U^c$ is contained in J. If $\overline{U} \cap U^c \neq J$, the let $A \subseteq J$ be an arc with $\overline{U} \cap U^c \subseteq A$. Now, since J is bounded, $\boldsymbol{R}^2 - J$ can have only one unbounded component. Thus, we must have at least one bounded component.

First, assume that U is a bounded component. Choose $a \in U$ and let D be a disk centred at a containing J. By the Tietze Extension theorem, the function $id : A \rightarrow A$ has a continuous extension $R : D \rightarrow A$. We define $f : D \rightarrow D \backslash \{a\}$ by

$$f(x) = \begin{cases} r(x) & ,if\ x \in \overline{U} \\ x & ,if\ x \in U^c \end{cases}$$

Note that $\overline{U} \cap U^c \subseteq A$. So, f is well-defined and continuous. Now, let $p : D \backslash \{a\} \rightarrow S$ be the projection through a onto the boundary S of the disk and $g : S \rightarrow S$ the antipodal map. Then, the function $g \circ f$

: $D \to D$ is continuous with no fixed point violating Brouwer's theorem. Thus, $\overline{U} \cap U^c = J$.

The case where U is unbounded follows similarly by letting

$$f(x) = \begin{cases} x & ,\text{if } x \in \overline{U} \\ r(x) & ,\text{if } x \in U^c \end{cases}$$

Next, Maehara uses Brouwer's theorem to prove the following. Suppose in a rectangle we have two paths, one going from the left to the right side and the other starting at the top and ending somewhere along the bottom. Then, these paths must intersect.

Theorem: *Let $f(t) = (f_1(t), f_2(t))$ and $g(t) = (g_1(t), g_2(t))$ for $-1 \leq t \leq 1$, be paths in $E(a, b; c, d)$ such that*

$$f_1(-1) = a, f_1(1) = b,\ g_2(-1) = c,\ g_2(1) = d.$$

Then, there exists $t \in [-1, 1]$ such that $f(t) = g(t)$.

Proof: Suppose the contrary. If $f(s) = g(t)$ for all $s, t \in [0, 1]$, then the function h defined by

$$h(s, t) = \frac{(g_1(t) - f_1(s), f_2(s) - g_2(t))}{\max\{|f_1(s) - g_1(t)|, |f_2(s) - g_2(t)|\}}$$

is well–defined and continuous and maps E into its boundary. Thus, by Brouwer's theorem, h has a fixed point say (s, t). But, $h(s, t) = (s, t)$ implies $|s| = 1$ or $|t| = 1$. Examining these four possibility gives a contradiction in each case.

Using these two results, Maehara proves the Jordan Curve theorem by first constructing a particular point $z \in \boldsymbol{R}^2 - J$ and proving that the component containing z must be bounded and then, showing there can be no other bounded component than this.

2.6.4 Existence of Equilibrium Points [*von Neumann*]

Kakutani's theorem has had far reaching influences in game theory and economic theory where the goal is to show the existence of an equilibrium point. There are literally hundreds of citations of the theorem in the literature. Famous results by Nash, von Neumann, and Ky Fan have been shown to be consequences of Kakutani's fixed point theorem. Here, we present a famous lemma by von Neumann, from which he obtained a fundamental result in the theory of games.

***Lemma* [von Neumann]:** *Let X and Y be nonempty compact convex subsets of* $\boldsymbol{R}^n$. *Suppose A and B are closed subsets of* $X \times Y$ *such that* $A(y) = \{x \in X : (x, y) \in A\}$ *and* $B(x) = \{y \in Y : (x, y) \in B\}$ *are nonempty closed convex subsets of X and Y respectively. Then,* $A \cap B$ *is nonempty and compact.*

The original proof by Von Neumann was somewhat complicated and used a notion of integrals in Euclidean space. The theorem is very easy to prove using Kakutani's theorem. In fact, Kakutani published his theorem as a method for proving this lemma from which it is not hard to obtain von Neumann's minimax theorem.

Proof: Define $F : X \times Y \to X \times Y$ by $F(x, y) = A(y) \times B(x)$. Compactness, convexity and non emptiness of $F(x, y)$ follow from that of its components parts. Also, it follows from the closedness of A and B that F is *usc*. Thus, by Kakutani's theorem, F has a fixed point, say $(x, y) \in A(y) \times B(x)$. It follows that $(x, y) \in A$.

From the lemma, it is not hard to obtain the famous von Neumann minimax theorem. We state the theorem here without proof.

Theorem : *Let* $K \subseteq \boldsymbol{R}^m$, $L \subseteq \boldsymbol{R}^n$ *be closed and convex and* $f : K \times L \to \boldsymbol{R}$ *continuous. Suppose for all* $x \in K$ *and* α *in* $\boldsymbol{R}$, $\{y \in L : f(x, y) \leq \alpha\}$ *is conve and for all* $y \in L$ *and* β *in* $\boldsymbol{R}$, $\{x \in K : f(x, y) \geq \beta\}$ *is convex. Then,*

$$\max_{x\in K} \min_{y\in L} f(x, y) = \min_{y\in L} \max_{x\in K} f(x, y).$$

2.6.5 The Fundamental theorem of algebra

In a 1949 paper, B.H. Arnold wrote "…it has been known for some time that the undamental theorem of algebra could be derived from Brouwer's fixed point theorem". He continued to give a simple one page proof that a polynomial with degree $n \geq 1$ has at least one complex root. Unfortunately, he did this by applying Brouwer's theorem to a discontinuous function. This error was spotted and a correction was published less thantwo years later . In 1951, M.K. Fort followed up with a brief proof using the existence of continuous *n*th roots of a continuous non zero function on a disk in the complex planeto show that Brouwer can be used to prove the fundamental theorem. In this author'sopinion, except for an ill–fated choice of titles, this would have probably been the end ofthe incorrect proof by Arnold. Arnold's paper was boldly called "A Topological Proof ofthe Fundamental Theorem of Algebra", whereas Fort's paper was modestly titled "Some Properties of Continuous Functions". Since 1949, citations of Arnold's paper have popped up from time to time as proof of the existence of a topological proof of the fundamental theorem of algebra, including in a four hundred page volume on fixed point theory published in 1981, a talk given in 2000 at a meeting of the Association for Symbolic Logic as well as in newsgroup discussions that are at least as recent as 1998. On the other hand, this author had a difficult time tracking down Fort's paper, even knowing before hand that it did exist. No amount of searching electronic databases with relevant key words would produce it though Arnold's paper was invariably returned. Perhaps, it would have been useful for Fort to rename his work "A Correct Topological Proof of the Fundamental Theorem of Algebra". It seems certain that Arnold would have preferred this to having his blunder quoted so longafter he made his apologies.

2.6.6 Haar measures

Topological groups: *A topological group is a group G endowed with a Hausdorff topology that makes the group operations continuous; namely, the map* $(x, t) \mapsto xy^{-1}$ *is continuous for every* $x, y \in G$.

For any $y \in G$ the maps $x \mapsto xy$, $x \mapsto yx$, and $x \mapsto x^{-1}$ are homeomorphisms of G onto G. Hence, the topology of G is uniquely determined by any local base at the identity element e. Indeed, if U is a neighborhood of some $x \in G$, the sets $xU = \{xy : y \in U\}$ and $Ux = \{yx : y \in U\}$ are neighborhoods of e.

The next topological lemma will be used in the proof of the main result of the section.

Definition: *Let G be a compact topological group. A Haar measure on G is a regular*

Lemma: *Let G be a compact topological group and let $K_1, K_2 \subset G$ be disjoint compact sets. Then, there exists an open neighborhood U of e such that no translate of U meets both K_1 and K_2.*

Borel probability measure μ which is simultaneously left invariant, i.e.,

$$\int_G f(x)\, d\mu(x) = \int_G f(yx)\, d\mu(x),\ \forall\ y \in G, f \in C(G)\ldots\ldots..(a)$$

and right invariant, i.e.,

$$\int_G f(x)\, d\mu(x) = \int_G f(xy)\, d\mu(x),\ \forall\ y \in G, f \in C(G)\ldots\ldots (b)$$

and satisfies the relation

$$\int_G f(x)\, d\mu(x) = \int_G f(x^{-1})\, d\mu(x),\ \forall\ f \in C(G)\ldots\ldots\ldots\ldots\ldots (c)$$

It is readily seen that a Haar measure on G, if it exists, is unique. Indeed, if μ and υ are two Borel probability measures on G, with μ left invariant and υ right invariant, for every $f \in C(G)$ the Fubini theorem yields

$$\begin{aligned}\int_G f(x)\, d\mu(x) &= \int_G f(yx)\, d\mu(x) \\ &= \int_G \Big[\int_G f(yx)\, d\mu(x)\Big]\, d\upsilon(y) \\ &= \int_G \Big[\int_G f(yx)\, d(y)\Big] d\mu(x) \\ &= \int_G f(yx)\, d\upsilon(y) \\ &= \int_G f(y)\, d\upsilon(y)\end{aligned}$$

forcing the equality $\mu = \upsilon$.

Theorem *: Let G be a compact topological group. Then, there exists a unique Haar measure on G.*

Proof: By means of the above uniqueness argument, it is enough to prove the existence of two regular Borel probability measure μ and υ satisfying *(a)* and *(b)* respectively. Then, *(c)* follows at once for if μ is a Haar measure on G, so is υ defined by $d\upsilon\,(x) = d\mu(x^{-1})$.

We then proceed by proving the existence of μ (the proof for υ being the same). Let $\mathcal{K}$ be the family of compact subsets of G and $\mathcal{U}$ be the family of open subsets of G containning e. If $K \in \mathcal{K}$ and $U \in \mathcal{U}$, we can always cover K by a finite number of translates of U. We define the *covering number* $[K{:}U]$ of K by U the smallest number of translates of U required to cover K. For every $K \in \mathcal{K}$, we intro-duce the normalized covering ratio

$$\xi_K(U) = \frac{[K:U]}{[G:U]}, \qquad \forall\, U \in \mathcal{U}$$

It is immediate to verify that $0 \le \xi_K(U) \le 1$, $\xi_{K_1}(U) \le \xi_{K_2}(U)$ for $K_1 \subset K_2$ and $\xi_{xK}(U) = \xi_K(U)$ for all $x \in G$. Moreover, if K_1, $K_2 \in \mathcal{K}$ are disjoint sets exploiting the lemma we find $U \in \mathcal{U}$ such that no translate of U meets both K_1 and K_2. Hence, any covering of $K_1 \cup K_2$ by translates of U is the disjoint union of coverings of K_1 and K_2. The same holds replacing U with any $V \in \mathcal{U}$, $V \subset U$. Thus,

$$\xi_{k_1 \cup k_2}(V) = \xi_{K_1}(V) + \xi_{K_2}(V), \quad \forall\ V \in \mathcal{U},\ V \subset U \ldots.(d)$$

Notice that $(\mathcal{U}, \cap)$ is an abelian semigroup and we know a theorem "if S be an abelian semigroup, if S is amenable, so this theorem applies yielding the existence of an invariant mean Λ on $\mathcal{U}$. Since $\xi_K : \mathcal{U} \mapsto \boldsymbol{R}$ belongs to $\ell^\infty(\mathcal{U})$, we can define $\psi\ (K) = \Lambda\, \xi_K$, $\forall\ K \in \mathcal{K}$.

The following hold:

$$\psi\ (\phi) = 0\,, \quad \psi\ (G) = 1; \ldots\ldots\ldots\ldots\ldots\ldots\ldots\ldots\ldots\ldots\ldots(e)$$
$$\psi\ (K_1) \leq \psi\ (K_2)\,,\ if\ \ K_1 \subset K_2 \ldots\ldots\ldots\ldots\ldots\ldots\ldots\ldots(f)$$
$$\psi\ (xK) = \psi\,(K),\ \forall\, x \in G \ldots\ldots\ldots\ldots\ldots\ldots\ldots\ldots\ldots\ldots(g)$$
$$\psi\,(K_1 \cup K_2) = \psi\,(K_1) + \psi\,(K_2),\ if\, K_1 \cap K_2 = \phi \ldots\ldots\ (h)$$

Properties (e) - (g) are direct consequences of the definition of $\xi_K(U)$ whereas (h) follows from (d) and the fact that if $\xi \in \ell^\infty(\mathcal{U})$ and there is $U \in \mathcal{U}$ such that $\xi\,(V) = 0$ for every $V \in \mathcal{U}$, $V \subset U$, then $\Lambda\xi = 0$. Indeed, $\Lambda\,(R_U\,\xi) = \Lambda\xi$, and $(R_U\,\xi)(W) = \xi\,(U \cap W) = 0$ for every $W \in \mathcal{U}$.

For each open set $O \subset G$ define

$$\mu^*(O) = \sup\{\psi\,(K) : K \subset O,\ K \text{ compact}\}$$

and for every $E \subset G$ let

$$\mu^*(E) = \inf\{\mu^*(O) : O \supset E,\ O \text{ open}\}.$$

By (e) - (g), $\mu^*(\phi) = 0$, $\mu^*(G) = 1$, $\mu^*(E_1) \leq \mu^*(E_2)$ if $E_1 \subset E_2 \subset G$ and μ^* is left invariant. Moreover, μ^* is coun-tably subadditive, i.e.,

$$\mu^*\left(\bigcup_{j=1}^{\infty} E_j\right) \leq \sum_{j=1}^{\infty} \mu^*(E_j) \ldots\ldots\ldots\ldots\ldots\ldots(i)$$

To see that, let $\varepsilon > 0$ and choose open sets $O_j \subset E_j$ with $\mu^*(O_j) \leq \mu^*(E_j) + \frac{\varepsilon}{2^j}$ and a compact set $K \subset \bigcup_{j=1}^{\infty} O_j$. Due to compactness, $K \subset \bigcup_{j=1}^{n} O_j$ for some n. Consider a partition of the unity $\varphi_1, \ldots, \varphi_n$ for K subordinate to the open cover $\{O_1, \ldots, O_n\}$, and set $K_j = \{x \in G : \varphi_j(x) > 1/n\}$. The K_j are compact, $\bigcup_{j=1}^{n} K_j \supset K$ and $K_j \cap K \subset O_j$. Thus,

$$\psi(K) \leq \sum_{j=1}^{n} \psi(K_j \cap k) \leq \sum_{j=1}^{n} \mu^*(O_j) \leq \sum_{j=1}^{n} \mu^*(E_j) + \varepsilon$$

Taking the supremum over all such K and letting $\varepsilon \to 0$, we get (i). We conclude that μ^* is an outer measure. Applying then the Carath´eodory extension process, we build a measure μ, which coincides with μ^* on the measurable sets, namely, those sets $E \subset G$ such that

$$\mu^*(T) = \mu^*(T \cap E) + \mu^*(T \cap E^C \ EC), \ \forall T \subset G.$$

The proof is finished if we show that the open sets are measurable since μ is then a Borel outer regular measure (by construction) and hence, regular (being finite). So, let O be an open set and T be any set. Let $\varepsilon > 0$ and take an open set $A \supset T$ such that $\mu^*(A) \leq \mu^*(T) + \varepsilon$. Choose a com- pact set $K \subset A \cap O$ with $\mu^*(A \cap O) \leq \psi(K) + \varepsilon$, and a compact set $K_0 \subset A \cap K^C$ with $\mu^*(A \cap K^C) \leq \psi(K_0) + \varepsilon$. Then,

$$
\begin{aligned}
&\mu^*(T\cap O) + \mu^*(T\cap O^C) \\
&\le \mu^*(A\cap O) + \mu^*(A \cap K^C) \\
&\le \psi(K) + \psi(K_0) + 2\varepsilon \\
&= \psi(K \cup K_0) + 2\varepsilon \\
&\le \mu^*(A) + 2\varepsilon \\
&\le \mu^*(T) + 3\varepsilon .
\end{aligned}
$$

Since ε is arbitrary, the desired conclusion is immediate.

A different proof of the above theorem, that directly applies the Markov-Kakutani theorem exists. The advantage of the approach presented here is that it can be easily modified to extend the result to the locally compact case.

2.6.7 Existence of Nash Equilibrium

Consider a two-person finite game in strategic form. Label the strategies of player 1 by 1, 2,…,m and label player two's strategies by 1, 2,…,n. Define the k-simplex by $\Delta_k = \{x \in \boldsymbol{R}_+^{k+1} : \sum_{i=1}^{k+1} x_i = 1\}$. We can regard any mixed strategy of player one as a point Δ_{m-1} and any mixed strategy of player two as a point in Δ_{n-1}. The payoff functions are defined on $S = \Delta_{m-1} \times \Delta_{n-1}$ which is the product of the strategy sets.

Given strategies $p \in \Delta_{m-1}$ and $q \in \Delta_{n-1}$, the expected pay off are

$$Eu_\iota(p,q) = \sum_{i=1}^{m}\sum_{j=1}^{n} p_i q_j u_\iota(i, j).$$

We define matrices A and B by $aij = u_1(i, j)$ and $b_{ij} = u_2(i, j)$. The expected payoffs can then be written. $Eu_1(p, q) = p'Aq$ and $Eu_2(p, q) = p'Bq$ where p and q are regarded as column vectors and the prime denotes transpose.

We will denote the ith column of A by A_i and the jth row of B by $B_{.j}$ Thus, Aiq gives the expected payoff to player one when playing the pure strategy i against player two's mixed strategy q.
Similarly, $p'B_j$ gives the expected payoff to player two when playing the pure strategy j against player one's mixed strategy p.

Theorem : *Every two-person finite game has a Nash equilibrium in mixed strategies.*

Proof: Define $c_i(p, q) = \max\{A_iq - p'Aq, 0\}$ for $i = 1,\ldots,m$ and $d_j(p, q) = \max\{p'B_j - p'Bq, 0\}$ for $j = 1,\ldots,n$. Thus, c_i and d_j represent the gain (if any) from switching to the pure strategy i (j) from the mixed strategy p (q).
Now define functions

$$P_i(p, q) = \frac{p_i + c_i(p,q)}{1 + \sum_{k=1}^{m} c_k(p,q)}$$

$$Qj(p, q) = \frac{q_j + d_j(p,q)}{1 + \sum_{k=1}^{n} d_j(p,q)}$$

Since $A_iq - p'Aq$ and $p'B_j - p'Bq$ are linear functions, they are continuous. Moreover, the maximum of two continuous functions is continuous which implies both c_i and d_j are continuous. Finally, both numerator and denominator of P_i and Q_i are continuous and the denominator is strictly positive. It follows that P_i and Q_i are continuous.

Furthermore,

$$\sum_{i=1}^{m} P_i(p, q) = \frac{\sum_{i=1}^{m} p_i + \sum_{i=1}^{m} c_i(p,q)}{1 + \sum_{k=1}^{m} c_k(p,q)}$$

$$= \frac{1+\sum_{k=1}^{m} c_k(p,q)}{1+\sum_{k=1}^{m} c_k(p,q)}$$

$$= 1$$

which means $P \in \Delta_{m-1}$. Similarly, $\sum_{j=1}^{n} Q_j(p, q) = 1$ and so $Q \in \Delta_{n-1}$.

Define the function T,

$$T: \Delta_{m-1} \times \Delta_{n-1} \rightarrow \Delta_{m-1} \times \Delta_{n-1} \text{ by}$$

$$T(p, q) = (P(p, q), Q(p, q)).$$

Note that T is continuous.

Moreover, $S = \Delta_{m-1} \times \Delta_{n-1}$ is compact and convex. The Brouwer Fixed Point Theorem yields $(p^*, q^*) \in \Delta_{m-1} \times \Delta_{n-1}$ such that $T(p^*, q^*) = (p^*, q^*)$. Let $u_1^* = p^{*\prime} A_{q^*}$ be player one's expected utility at (p^*, q^*).

Claim: $\sum_{k=1}^{m} c_k(p^*, q^*) = 0$.

Suppose the claim is false. Then, $\sum_{k=1}^{m} c_k(p^*, q^*) > 0$. Because, (p^*, q^*) is a fixed point,

$$p_i^* = \frac{p_i^* + c_i(p^*, q^*)}{1+\sum_{k=1}^{m} c_k i(p^*, q^*)}$$

for every i. Clearing the fraction and cancelling p_i^* yields $p_i^*[\sum_k c_k(p^*, q^*)] = c_i(p^*, q^*)$. Thus, $p_i^* = 0$ whenever $c_i(p^*, q^*) = 0$.

Let $I = \{i : p_i^* > 0\}$. Note that $I \neq 0$; because $\Sigma_i\, p_i^* = 1$. By the previous paragraph, $I \subset \{i : c_i(p^*, q^*) > 0\}$. Finally, the definition of I implies

$$\sum_{i=1}^{m} p_i^* = \Sigma_{i \in I}\, p_i^* = 1.$$

For $i \in I$, $c_i(p^*, q^*) > 0$ which means $A_i q^* > u_1^*$. Multiplying by $p_i^* > 0$ yields $p_i^* A_i q^* > p_i^* u_1^*$. Summing

$$u_1^* = \sum_{i=1}^{m} p_i^* A_i q^*$$

$$\geq \Sigma_{i \in I}\, p_i^* A_i q^*$$

$$> (\Sigma_{i \in I}\, p_i^*) u_1^* = u_1^*.$$

But this is a contradiction, which means our supposition that $\sum_{k=1}^{m} c_k(p^*, q^*) > 0$ must be false. This establishes our claim.

Since $= \sum_{k=1}^{m} c_k(p^*, q^*)\ 0$, $c_i(p^*, q^*) = 0$ for every $i = 1,\ldots,m$. expanding c_i yields $A_i q^* \leq u_1^*$.

Let $p \in \Delta_{m-1}$. Then,

$$p' A q^* = \sum_{i=1}^{m} p_i^* A_i q^*$$

$$\leq (\Sigma_{i \in I}\, p_i^*) u_1^* = u_1^*.$$

In other words, p^* is a best response to q^*. A similar argument shows that q^* is a best response to p^*. Since p^* and q^* are mutual best responses, we have a Nash equilibrium.

Chapter 3
Metric Fixed Point Theory

3.1 Introdution

Metric fixed point theory deals with the branch of fixed point theory which metric conditions on the underlying space and/or on the mappings play a fundamental role. It lies in the "happy medium" between the more general topological soace and the restrictive Bnach space. A natural extension of the study of contractions is the limiting case when the Lipschitz constant is allowed to equal one. Such mapings are called *nonexpansive*.

Because, most of the spaces studied in analysis share many algebraic and topological properties as well as metric properties, there is no clear line separating metric fixed point theory from the topological or set-theoretic branch of the theory. Also, metric fixed point theory has provided the motivation for the study of many geometric properties of Banach spaces

3.2 Banach's Contraction Principle, extensions and converse

3.2.1 Introduction

In 1910, after Brouwer gave the first fixed point theorem (although Bolzano's Intermediate Value Theorem of 1817 is the first, that of Brouwer is considered to be the first), the interest in the field of fixed point grew more and more.

3.2.2 Banach's Contraction Principle

Denition : *Let* (M, d) *be a metric space. The map* $T : M \to M$ *is said to be lipschitzian if there exists a constant* $k > 0$ *(called lipschitz constant) such that*

$$d(T(x), T(y)) \leq k\, d(x, y) \text{ for all } x, y \in M.$$

A lipschitzian mapping with a lipschitz constant k less than 1, i.e. $k < 1$ is called contraction and k is called the contraction modulus of T.

Theorem A : *Every contraction mapping is continuous.*

Proof: Let $T : X \to X$ *be* a contraction on a metric space (X, d) with modulus k and let $y \in X$.

Let $\varepsilon > 0$ and let $\delta = \varepsilon$. Then, $d(x,y) < \delta \Rightarrow d(Tx,Ty) \leq k\delta < \varepsilon$. Therefore, T is continuous at y. Since y was arbitrary, T is continuous on X.

The above proof actually establishes that a contraction mapping is uniformly continuous.

Definition: *Let (X, d_X) and (Y, d_Y) be metric spaces. A function $f : X \to Y$ is uniformly continuous if for every $\varepsilon > 0$ there is a $\delta > 0$ such that*

$$\forall\; x, y \in X : d_X(x, y) < \delta \Rightarrow d_Y(f(x), f(y)) < \varepsilon$$

Notice how this definition differs from the definition of continuity, uniform continuity requires that, for a given ε, a single δ will work across the entire domain of f. Continuity allows that the δ may depend upon the point x at which continuity of f is being evaluated; uniform continuity requires that (for a given ε) being within δ of any $x \in X$ guarantees that the image under f is within ε of $f(x)$.

***Theorem B*:** *Every contraction mapping is uniformly continuous.*

***Theorem C* [*Banach's Contraction Principle*] :** *Every contraction mapping on a complete metric space has a unique fixed point*. (This is also known as Contraction Mapping Theorem or The Banach's Fixed Point Theorem)

***Proof*:** Let $T : M \to M$ be a contraction on the complete metric space (M, d) and let k be a contraction modulus of T. First, we show that T can have at most one fixed point. Then, we construct a sequence which converges and show that its limit is a fixed point of T.

(*a*) Suppose x and y are fixed points of T. Then, $d(x, y) = d(Tx,Ty) \leq kd(x, y)$; since $k < 1$, this implies that $d(x, y) = 0$, i.e., $x = y$.

(*b*) Let $x_0 \in X$ and define a sequence $\{x_n\}$ as follows:

$$x_1 = Tx_0, \quad x_2 = Tx_1 = T^2x_0, \dots, \quad x_n = T\,x_{n-1} = T^n x_0, \dots$$

We first show that adjacent terms of $\{x_n\}$ grow arbitrarily close to one another, specially that $d(x_n, x_{n+1}) \leq k^n d(x_0, x_1)$:

$$\begin{aligned}
&d(x_1, x_2) \leq k\, d(x_0, x_1)\\
&d(x_2, x_3) \leq k\, d(x_1, x_2) \leq k^2\, d(x_0, x_1)\\
&\dots\dots\dots\dots\dots\dots\\
&\dots\dots\dots\dots\dots\dots\\
&d(x_n, x_{n+1}) \leq k\, d(x_{n-1}, x_n) \leq k^n d(x_0, x_1).
\end{aligned}$$

Next, we show that if $n < m$, then $d(x_n, x_m) < k^n \dfrac{1}{1-k} d(x_0,x_1)$:

$$\begin{aligned}
d(x_n, x_{n+1}) &\leq k^n d(x_0, x_1)\\
d(x_n,x_{n+2}) &\leq d(x_n, x_{n+1}) + d(x_{n+1}, x_{n+2})\\
&\leq k^n d(x_0, x_1) + k^{n+1}\, d(x_0, x_1)\\
&= (k^n + k^{n+1}) d(x_0, x_1)\\
&\dots\dots\dots\dots\dots\dots\dots\dots\\
&\dots\dots\dots\dots\dots\dots\dots\dots\\
d(x_n, x_m) &\leq (k^n + k^{n+1} + \dots + k^{m-1})\, d(x_0, x_1)\\
&= k^n(1 + k + k^2 + \dots + k^{m-1-n})\, d(x_0, x_1)\\
&< k^n(1 + k + k^2 + \dots)\, d(x_0, x_1)\\
&= k^n \frac{1}{1-k}\, d(x_0, x_1).
\end{aligned}$$

Therefore, $\{ x_n \}$ is Cauchy, i.e., for $\varepsilon > 0$, let N be large enough

$k^N \dfrac{1}{1-k} d(x_0, x_1)$ which ensures that $n, m > N \Rightarrow d(x_n, x_m) < \varepsilon$. Since the metric space (M, d) is complete, the Cauchy sequence $\{x_n\}$ converges to a point $m \in M$. We show that m is a fixed point of T. Since $x_n \to m$ and T is continuous, we have $Tx_n \to Tm$, i.e., $x_{n+1} \to Tm$. Since $x_{n+1} \to m$ and $x_{n+1} \to Tm$, we have $Tm = m$.

Corollary D **:** *Let M be a complete metric space and Y be a topological space. Let $T : M \times Y \to M$ be a continuous function and T be a contraction on M uniformly in Y, that is,*

$$d(T(x_1, y), T(x_2, y)) \le \lambda\, d(x_1, x_2),\ \forall\ x_1, x_2 \in M,\ \forall\ y \in Y,$$

for some $\lambda < 1$, then, for every fixed $y \in Y$, the map $x \mapsto T(x, y)$ has a unique fixed point $\varphi(y)$. Moreover, the function $y \mapsto \varphi(y)$ is continuous from Y to M.

Corollary E **:** *Let M be a complete metric space and let $T{:}M \to M$. If T^n is a contraction for some $n \ge 1$, then T has a unique fixed point $m^* \in M$.*

Remark *: When T: $M \to M$ is a contraction with constant k, any iterate T^n is a contraction with constant k^n, the unique fixed point of T will also be the unique fixed point of any T^n.*

3.2.3 Further extensions of the contraction principle

3.2.3(a) Caristi-Ekeland Principle

This is one of the most interesting extensions of Banach's Contraction Principle.

Let $f : M \to \mathbf{R}$ and assume that f is lower semicontinuous (l.s.c.), i.e. for any sequence $\{x_n\} \subset M$, if $\lim_{n\to\infty} x_n = x$ and $\lim_{n\to\infty} f(x_n) = r$, then $f(x) \le r$.

***Theorem* [*Ekeland, 1974*] :** *Let (M, d) be a complete metric space and $f : M \to \mathbf{R}^+$ l.s.c. Define:*

$$x \le y \Leftrightarrow d(x, y) \le f(x) - f(y), \quad x,\ y \in M.$$

Then, $(M, \le)$ *has a maximal element.*

***Theorem* [*Caristi, 1975*] :** *Let M and f be as above. Suppose g : M → M satisfies:*

$$d(x, g(x)) \le f(x) - f(g(x), \qquad x \in M$$

Then, g has a fixed point.

The proof of is based on discrete technique: Zorn's lemma and Axiom of Choice applied to Bronsted partial order.

3.2.3(b) Boyd-Wong Theorem

***Theorem* [*Boyd-Wong*] :** *Let M be a complete metric space and let* $T{:}M \to M$. *Assume there exists a right-continuous function* $\varphi : [0, \infty) \to [0, \infty)$ *such that* $\varphi(r) < r$ *if* $r > 0$ *and*

$$d(M(x_1), T(x_2)) \le \varphi(d(x_1, x_2)), \qquad \forall\ x_1, x_2 \in M.$$

Then, T has a unique fixed point $m^* \in M$. *Moreover, for any* $m_0 \in M$ *the sequence* $T^n(m_0)$ *converges to* m^*.

3.2.3(c) Ciric Theorem

***Theorem* [*Ciric*] :** *Let X be a complete metric space and let* $f : X \to X$ *be such that*

$$d(f(x_1), f(x_2)) \le \lambda \max\{d(x_1, x_2), d(x_1, f(x_1)), d(x_2, f(x_2)), \\ d(x_1, f(x_2)), d(x_2, f(x_1))\}$$

for some $\lambda < 1$ *and every* $x_1, x_2 \in X$. *Then, f has a unique fixed point* $x^* \in X$. *Moreover,* $d(f^n(x_0), x^*) = O(\lambda^n)$ *for any* $x_0 \in X$.

3.2.4 Generalization

In some situations, a function is not a contraction but an iterate of it is. This turns out to suffice to get the conclusion of the contraction mapping theorem for the original function.

Theorem *: If M is a complete metric space and* $T : M \to M$ *is a mapping such that some iterate* $T^N : M \to M$ *is a contraction, then T has a unique fixed point. Moreover, the fixed point of T can be obtained by iteration of T starting from any* $x_0 \in M$.

Proof: By the contraction mapping theorem, T^N has a unique fixed point. Call it a, so $T^N(m) = m$. To show m is the only possible fixed point of T, observe that a fixed point of T is a fixed point of T^N and thus, must be m. To show m really is a fixed point of T, we note that $T(m) = T(T^N(m)) = T^N(T(m))$. So, $T(m)$ is a fixed point of T^N. Therefore, $T(m)$ and m are both fixed points of T^N. Since T^N has a unique fixed point, $T(m) = m$.

We now show that for any $x_0 \in M$ the points $T^n(x_0)$ converge to m as $n \to \infty$. Consider the iterates $T^n(x_0)$ as n runs through a congruence class modulo N. That is, pick $0 \le r \le N-1$ and look at the points $T^{Np+r}(x_0)$ as $p \to \infty$. Since

$$T^{Np+r}(x_0) = T^{Np}(T^r(x_0)) = (T^N)^p(T^r(x_0)),$$

these points can be viewed (for each r) as iterates of T^N starting at the point $y_0 = T^r(x_0)$. Since T^N is a contraction, these iterates of T^N (from any initial point, such as y_0) must tend to m by the contraction mapping theorem. This limit is independent of the value of r in the range $\{1, \ldots, N-1\}$, so all N sequences $\{T^{Np+r}(x_0)\}_{p \ge 1}$ tend to m as $p \to \infty$. This shows $T^n(x_0) \to m$ as $n \to \infty$.

3.2.5 Converse of Banach's Contraction Principle

In other words, what kind of set M will have the conclusion of Banach Contraction Principle? The most elegant result in this direction is due to Bessaga.

Definition: *A metric space (X, d) is ultrametric if and only if for each $x, y, z \in M$, $d(x, y) \leq \max\{d(x, z), d(y, z)\}$*
Theorem : *Let (X, τ) be a T_1 topological space and let $f: X \to X$ be a function which has a unique fixed point a and is such that for each $x \in X$, the sequence $(f^n(x))$ converges to a in τ. Then, there exists a function $d: X \times X \to \boldsymbol{R}$ such that (X, d) is a complete ultrametric space and such that for all $x, y \in X$, we have*

$$d(f(x, f(y)) \leq \frac{1}{2} d(x, y).$$

Proof: The proof is divided into several steps, numbered consecutively.

***Step 1*:** Given $x \in X$, we define the set $T(x) \subseteq X$ to be the smallest subset of X which is closed under the following rules :
(1) $x \in T(x)$
(2) If $y \in T(x)$ and $f(y) \neq a$, then $f(y) \in T(x)$
(3) If $y \in T(x)$ and $y \neq a$, then $f^{-1}(y) \subseteq T(x)$.

It is clear that the intersection of the family of all sets closed under these rules is itself closed under these rules and so the minimality condition is met. Moreover , it is also clear that each of the sets $T(x)$ is non empty. Now, let $\mathscr{T} = \{T(x) : x \in X\}$ and observe the following facts.

(i) $T(a) = \{a\}$. To see this, we note that (1), (2) and (3) are all true relative to the set $\{a\}$. Therefore, by minimality, we have $T(a) = \{a\}$.
(ii) If $x \neq a$, then $a \notin T(x)$ and so $T(a) \cap T(x) = \phi$. Hence, either $T(a)$ and $T(x)$ are equal or they are disjoint. To see this, suppose $x \neq a$ and consider rule (3). Clearly, we cannot have $a \in f^{-1}(x)$ otherwise $f(a) = x$ which is a contradiction. Thus, rules (2) and (3) applied repeatedly and starting with x never place a in $T(x)$ and by minimality, the process just described generates T (x).
(iii) If $T(x) \neq T(a)$ and $T(y) \neq T(a)$, then either $T(x)$ and $T(y)$ are equal or the are disjoint. To see this, suppose $z \in T(y) \cap T(x)$.

Then, the rules (1), (2) and (3) under repeated application starting with z force $T(x) = T(y)$. Thus, the collection $\mathcal{T}$ is a patition of x.

Step 2 : We next inductively define a mapping $\ell : T \to \mathbf{Z} \cap \infty$ and each T $\in \mathcal{T}$.

(1) We let $\ell(a) = \infty$ and this defines ℓ *on* $T = T(a)$. If $T \neq T(a)$, we choose an arbitrary $x \in T$ and set $\ell(x) = 0$ (of course, $x \neq a$) and proceed as follows:

(2) For each $y \in T$ with $f(y) \neq a$ and $\ell(y) = k$, let $\ell(f(y)) = k+1$.

(3) For each $y \in T$ with $\ell(y) = k$, let $\ell(z) = k-1$ for all $z \in f^{-1}(y)$.

We will, henceforth, assume that all this is done for every T $\in \mathfrak{T}$ so that ℓ is a function defined on all of X. It is clear that the mapping ℓ is well defined since (X, τ) is a T_1 space. For, if there is a cycle in the sequence $f^n(x)$ of iterates for some $x \in X$, then we can arrange for some element y in the sequence to be frequently not in some neighborhood of a, using the fact that X is T_1, which contradicts the convergence of the sequence $f^n(x)$ to a.

Step 3 : Define a mapping $p : Z \cup \{\infty\} \to \boldsymbol{R}$ by

$$p(k) = \begin{cases} 0 & \textit{if } k = \infty \\ 2^{-k} & \text{otherwise} \end{cases}$$

Furthermore, define a mapping $q : X \times X \to \boldsymbol{R}$ by

$$q(x, y) = \max\{\, p(\ell(x)),\, p(\ell(y))\}$$

and a mapping $d : X \text{ x } X \to \boldsymbol{R}$ by

$$d(x, y) = \begin{cases} q(x, y) & \text{if } x \neq y \\ 0 & \textit{if } x = y \end{cases}$$

Step 4 **:** We show that (X, d) is an ultrametric space.
(i) Let $d(x, y) = 0$ and assume that $x \neq y$. Then, $q(x, y) = d(x, y) = 0$. Hence, $\max\{p(\ell(x)), p(\ell(y))\}=0$.
So, $p(\ell(x)) = p(\ell(y))= 0$. Hence, $\ell(x) = \ell(y) = \infty$ and $x= y = a$ by construction of ℓ, a contradiction.
(ii) This is true by definition of d.
(iii) This is true by symmetry of q and hence, of d.
(iv) Let $x, y, z \in X$ and assume, without loss of generality, that $p(\ell(x)) < p(\ell(z))$ so that $d(x, z) = p(\ell(z))$. If $p(\ell(y)) \leq p(\ell(z))$, then $d(y, z) = p(\ell(z))$. If $p(\ell(y)) > p(\ell(z))$, then $d(y, z) = p(\ell(y)) > p(\ell(z))$. In both cases, we get $d(y, z) \geq d(x, z)$ as required.

Step 5 **:** (X, d) is a complete metric space. In order to show this, let (x_n) be a Cauchy sequence in X. If (x_n) is eventually constant, then it converges trivially. So, assume that (x_n) is not eventually constant. We show that x_n converges to a in d for which it suffices to show that $(p(\ell(x_n)))_{n \in N}$ converges to 0. Let $\varepsilon > 0$. Then there exists $n_0 \in N$ such that for all $m, n \geq n_0$ we have $d(x_m, x_n) < \varepsilon$. In particular, we have $d(x_m, x_{n_0}) < \varepsilon$ for all $m \geq n_0$ and since (x_n) is not eventually constant, we obtain $p(\ell(x_{n_0})) < \varepsilon$ and also $p(\ell(x_m)) < \varepsilon$ for all $m \geq n_0$. Since ε was chosen arbitrarily, we see that $(p(\ell(x_n)))_{n \in N}$ converges to 0.

Step 6 **:** We note that for $f(x) \neq a$, we have $\ell(f(x)) = \ell(x) + 1$ by definition of ℓ and hence

$$p(\ell(f(x))) = \frac{1}{2}\, p(\ell(x)) .$$

Step 7 **:** For all $x, y \in X$, we have that $d(f(x), f(y)) \leq \frac{1}{2}\, d(x, y)$. In order to show this, let $x, y \in X$ and without loss of generality that $x \neq y$. Let $d(x, y)= 2^{-k}$, say, so that $\max\{p(\ell(x)), p(\ell(y))\} = 2^{-k}$. Then,

$$d(f(x), f(y)) = \max\{p(\ell(f(x))),\ p(\ell(f(y)))\}$$

$$= \frac{1}{2}\max\{p(\ell(x)), p(\ell(y))\}$$

$$= \frac{1}{2}d(x, y) \text{ as required.}$$

N.B : *It should be noted that this theorem is not a true converse of Banach's contraction principle. In this, we don't start with a metrizable space and attempt to obtain a metric for in relative to which f is a contraction.*

3.3 Metric Fixed Point Theory in Banach Spaces

3.3.1 Introduction

The formal definition of Banach spaces is due to Banach himself. But examples like the finite dimensional vector space $\boldsymbol{R}^n$ was studied prior to Banach's formal definition of Banach spaces. In 1912, Brouwer proved the following:

Theorem : *Let B be a closed ball in $\boldsymbol{R}^n$. Then any continuous mapping $T : B \to B$ has at least one fixed point.*

This theorem has a long history. The ideas used in its proof were known to Poincare as early as 1886. In 1909, Brouwer proved the theorem when $n = 3$. And in 1910 Hadamard gave the first proof for arbitrary n and Brouwer gave another proof in 1912. All of these are older results than the Banach Contraction Principle. Though in nature the two theorems are different, they bare some similarities. A combination of the two led to the so called metric fixed point theorem in Banach spaces. Indeed, in Brouwer's theorem the convexity, compactness and the continuity of T are crucial while the lipshitz behavior of the contraction and completeness are crucial in Banach's fixed point theorem. In infinite normed linear vector spaces, we lose the compactness of the bounded closed convex sets (like closed balls). So, if we assume completeness we get Banach spaces. On these, we have another natural topology (other than the norm topology) that is the weak topology. So, a weakly-compact convex set needs not to be compact for the norm. The best example here is the Hilbert space. In it any bounded closed convex set is weakly compact. The lipschitz condition to be considered is when the lipschitz constant is equal to 1. Such mappings are called *non- expansive*. In other words, if (M, d) is a metric space, then $T : M \to M$ is nonexpansive if $d(T(x), T(y)) \leq d(x; y)$ for any $x, y \in M$.

The metric fixed point problem in Banach spaces becomes:

Let X be a Banach space and C a nonempty bounded closed convex subset of X. When does any nonexpansive mapping $T : C \to C$ have at least one fixed point ?

Other interesting problems closely related to this one are:

1. The structure of the fixed points sets.
2. The approximation of fixed points.
3. Abstract Metric Theory.

Recognition of fixed point theory for nonexpansive mappings as a noteworthy avenue of research almost surely dates from the 1965 publication of

1. *Browder-Gohde Theorem:* *If K is a bounded closed convex subset of a uniformly convex Banach space X and if $T : K \to K$ is nonexpansive, then T has a fixed point. Moreover, the fixed point set of T is a closed convex subset of K.*

2. *Kirk Theorem:* *If K is a weakly-compact convex subset of a Banach space X, assume that K has the normal structure property, then any nonexpansive mapping $T : K \to K$ has a fixed point.*

3.3.2 Classical Existence Results

While studying a paper by Brodskii and Milman (1948), Kirk (1965) was able to discover the stated result. So the normal structure property is an old concept not directly related to nonexpansive mappings. Hilbert space and uniformly convex Banach spaces havethe normal structure property. In fact, the proof given indepenpendtly by Browder and Gohde of their results do not use this property.

Let K be a bounded closed convex subset of a Banach space X and $T : K \to K$ be a nonexpansive mapping. Let $\varepsilon \in (0, 1)$ and $x_0 \in K$. Define T_ε by

$$T_\varepsilon(x) = \varepsilon x_0 + (1 - \varepsilon) T(x), \quad \text{for any } x \in K.$$

It is easy to check that $T_\varepsilon(K) \subset K$ (since K is convex) and it is a contraction. The Banach's Contraction Principle implies the existence of a unique point $x_\varepsilon \in K$. K such that

$$x_\varepsilon = \varepsilon x_0 + (1 - \varepsilon) T(x_\varepsilon)$$

which implies

$$\| x_\varepsilon - T(x_\varepsilon) \| = \varepsilon \| x_0 - T(x_\varepsilon) \| \leq \varepsilon \operatorname{diam}(K)$$

Therefore, we have

$$\inf_{x \in K} \| x - T(x) \| = 0$$

This property is known as *the approximate fixed point property*. So, any nonexpansive mapping defined on a bounded closed convex subset of Banach space has this property. By taking a sequence $\{\varepsilon_n\}$ which goes to 0, we generate a sequence of points $\{x_n\}$ from K such that

$$\lim \| x_n - T(x_n) \| = 0.$$

Such sequence is called an *approximate fixed point sequence (a.f.p.s.)*.

In order to grasp the ideas behind the proofs of the existence of fixed point for non-expansive mappings in the Hilbert and uniformly convex spaces, we need the concept of *asymptotic center* of a sequence discovered by Edelstein. This concept is very useful whenever one deals with sequential approximations in Banach spaces.

Let $\{x_n\}$ be a bounded sequence in a Banach space X and let C be a closed convex subset of X. Consider the functional $f : C \to [0, \infty)$ defined by

$$f(x) = \limsup ||x_n - x||$$

Usually we use the notation $f(x) = r(x, \{x_n\})$. The infimum of $f(x)$ over C is called the *asymptotic radius* of $\{x_n\}$ and denoted by $r(C, \{x_n\})$,i.e.

$$r(C, \{x_n\}) = \inf_{x\in C} r(x, \{x_n\}) = \inf_{x\in C} (\limsup|| x_n - x||)$$

The set

$$A(C, \{x_n\}) = \{x \in C ; r(x, \{x_n\}) = r(C, \{x_n\})\}$$

is the set of all asymptotic centers of $\{x_n\}$. If $\{x_n\}$ converges to $x \in C$, then $A(C, \{x_n\}) = \{x\}$. In general the set $A(C, \{x_n\})$ is not reduced to one point.

Theorem: *Every bounded sequence in a uniformly convex Banach space X has a unique asymptotic center with respect to any closed convex subset of X.*

In the Hilbert case, we get more:

Theorem: *In a Hilbert space H, the weak limit of a weakly convergent sequence coincides with its asymptotic center with respect to H.*

Note that in any Hilbert space H, the asymptotic center of a bounded sequence $\{x_n\}$ with respect to H belongs to the closed convex hull of $\{x_n\}$. But this does not necessarily hold in (even uniformly convex) Banach spaces.

Let C be a bounded closed convex subset of a uniformly convex Banach space X and $T : C \to C$ be nonexpansive.
1. Let $\{x_n\}$ be an *a.f.p.s.* of T in C. Let $z \in C$ be its asymptotic center with respect to C, then it is quite easy to check that $T(z)$ is also an asymptotic center of $\{x_n\}$ with respect to C. By the uniqueness of the asymptotic center, we get $T(z) = z$.

2. Let $x \in X$ and consider the orbit of x under T, i.e. $\{T^n(x)\}$. Let $z \in C$ be its asymptotic center with respect to C, then again it is quite easy to check that $T(z)$ is also an asymptotic center of $\{T^n(x)\}$ with respect to C. By the uniqueness of the asymptotic center, we get $T(z) = z$.

Remark: *In 1965, Browder in fact discovered something truly amazing. Let C be a bounded closed convex subset of a uniformly convex Banach space X and $T : C \to C$ be nonexpansive. If $\{x_n\} \subset C$ converges weakly to x and $\{x_n - T(x_n)\}$ converges strongly (with respect to the norm) to 0, then we have*

$$x - T(x) = 0$$

This is known as the Demiclosedness principle.

As we mentioned before, Kirk's theorem is based on the normal structure property discovered by Brodskii and Milman.

3.3.3 The Normal Structure Property

The reason behind separating it from the other results is the important role it played during the first 20 years (since 1965) of the theory. In order to appreciate this property, let us give more information on non-expansive mappings in Banach spaces. Indeed, let C be a weakly compact convex subset of a Banach space X. Let $T : C \to C$ be non-expansive. The first pioneers of the metric fixed point theory were mostly concerned about the existence of fixed points. So, assume that T fails to have a fixed point in C. Since C is weakly compact, there exists (by Zorn) a minimal nonempty closed convex subset K of C invariant under T, i.e. $T(K) \subset K$. The central research of the metric fixed point theory in Banach spaces always centered around the discovery of new properties of such minimal convex sets. The first property is due to Kirk.

Theorem: *Under the above notations, we have*

1. $\overline{conv}(T(K)) = K$.
2. $sup_{x \in K} ||z - x|| = diam(K) > 0,$ *for any* $z \in K$.

The normal structure property forbids the conclusion 2 to hold.

Definition : *A closed convex subset C of a Banach space X is said to have the normal structure property if any bounded convex subset K of C which contains more than one point, contains a nondiametral point, i.e. there exists a point* $x_0 \in K$ *such that*

$$\sup_{x \in K} ||x_0 - x|| < \mathrm{diam}(K)$$

We will also say that X has the normal structure property if any bounded closed convex subset has the normal structure property.

Throughout this section, we will use the following notations:

$$\mathrm{diam}(C) = \sup_{x,\, y \in C} ||x - y||$$

$$r_z(C) = \sup_{x \in C} ||z - x||$$

$$r(C) = \inf_{x \in C} r_x(C);$$

$$\mathcal{C}(C) = \{\, z \in C;\, r_z(C) = r(C) \,\}$$

The number $r(C)$ and the set $\mathcal{C}(C)$ are called the *Chebyshev radius* and *Chebyshev center* of C respectively.

Now we are ready to state Kirk's theorem.

Theorem: *Let X be a Banach space and suppose that C is a nonempty weakly compact convex subset of X which has the normal structure property. Then, any nonexpansive mapping* $T : C \to C$ *has a fixed point.*

For the following ten years most of the results were solely concentrated on the study of the normal structure property. For example, most of the classical known Banach spaces have normal structure property. So, most of the fixed point existence theorems were in fact theorems about normal structure property. It got so bad that some thought that any reflexive (or superreflexive) Banach spaces have the normal structure property. It was not before 1972 (1968??) that James renormed the Hilbert space l_2 to get rid of this property.

For any $\beta > 0$, define the new norm $|| \, . \, ||_\beta$ on l_2 by

$$|| \, . \, ||_\beta = \max \{||x||_2 , \ \beta \ |||x||_\infty \}$$

Set $X_\beta = (l_2, || \, . \, ||_\beta)$. James showed that the supereflexive Banach space $X_{\sqrt{2}}$ fails the normal structure property. This answered the question asked by the people working on the fixed point property in Banach spaces. Right after James published his result the natural question of whether $X_{\sqrt{2}}$ has the fixed point property arose. It was Karlovitz (in 1975) who settled this question to the affirmative. We will discuss more Karlovitz ideas later on. Hence, we see that normal structure property (which is a geometric property) implies the fixed point property but they are not equivalent. Then, people started to dissociate themselves from the normal structure property. In fact, Baillon and Schoneberg (1981) studied the space $X_{\sqrt{2}}$ and proved that this space has a geometric, known as *asymptotic normal structure property* which implies the fixed point property. In fact, they proved that X_β has this geometric property for $\beta < 2$. Then, it was asked what happened to X_β for $\beta \geq 2$. No definitive answer was given to a point that some though again that may be these spaces may lead an example of a superreflexive space which fails the fixed point property. No luck again. This time it was Lin (1985) who settled this question by proving that X_β has the fixed point property for any $\beta > 0$. After this and other results some went too far to claim that no geometric property is equivalent to the fixed point property. This question is still open.

3.3.4 More on Normal Structure Property

Since Kirk publication of his paper, people intensified their investigation of this property. For example, it was proved that uniformly convex Banach spaces have this property. In particular, it was shown that if X is a Banach space set

$$\delta_X(\varepsilon) = \inf \left\{ 1 - \left\| \frac{x+y}{2} \right\| : \|x\| \leq 1 , \|y\| \leq 1 , \|x-y\| \geq \varepsilon \right\}$$

defined for $\varepsilon \in [0, 2]$. The characteristic (or coefficient) of convexity of the Banach space X is the number

$$\varepsilon_0(X) = \sup \{ \varepsilon \geq 0 : \delta(\varepsilon) = 0 \}$$

X is *uniformly convex* if and only if $\varepsilon_0(X) = 0$, i.e. $\delta(\varepsilon) > 0$ for any $\varepsilon \in [0, 2]$. And X is said to be *uniformly non-square* if and only if $\varepsilon_0(X) < 2$. James studied these spa- ces extensively.

Assume that $\varepsilon_0(X) < 1$ and let $\varepsilon \in (\varepsilon_0(X), 1)$. Then, for any nonempty bounded closed convex subset of X, there exists a point $x \in C$ such that

$$r(x,C) \leq (1 - \delta(\varepsilon))\, \text{diam}(C)$$

So, we have

$$N(X) = \sup \left\{ \frac{r(C)}{diam(C)} \right\} \leq (1 - \delta(\varepsilon_0(X)))$$

where the supremum is taken over all nonempty bounded closed convex subset of X with more than one point. The number $N(X)$ is known as *the coefficient of uniform normal structure property*. So, we have the normal structure property and even more: uniform normal structure property, i.e. $N(X) < 1$. So, uniformly convex Banach spaces have a stronger geometric property. In fact, a weaker version was discovered to still imply the normal structure property. It is known as *uniform convexity in every direction* (*U.C.E.D.*). Indeed, let $z \in X$ be a unit vector ($||z|| = 1$). The *modulus of uniform convexity* $\delta_X(\varepsilon, z)$ of X in the direction z is defined by

$$\inf \left\{ 1 - \left\| \frac{x + y}{2} \right\| \; : \; \|x\| \le 1 \, , \, \|y\| \le 1 \, , x - y = \alpha z \, , \, \|x - y\| \ge \varepsilon \right\}$$

It is not hard to show that if X is *U.C.E.D.* Then, for any nonempty bounded closed convex subset K of X, the Chebishev center $C(K)$ of K has at most one point. So, clearly if X is *U.C.E.D.*, then it has the normal structure property. Spaces which are *U.C.E.D.* or equivalently renormed to be *U.C.E.D.* were extensively studied by Zizler (1971).

Remark : *There is a sequential characterization of the normal structure property discovered by Brodskii and Milman in their original work which played a key role in studying this property.*

A bounded sequence $\{x_n\}$ in a Banach space is said to be a diametral sequence if

$$lim\ dist\ (_{n+1}, \overline{conv}\ \{x_1, x_2, \ldots, x_n\}) = diam\{x_1, x_2, \ldots.\}$$

We have

Theorem : *A bounded convex subset K of a Banach space has normal structure if and only if it does not contain a diametral sequence.*

3.3.5 Normal Structure and Smoothness

As with convexity, it is possible to scale Banach spaces with respect to their smoothness.

Let X be a Banach space. Define the *modulus of smoothness* ρ_X of X by

$$\rho_X(\eta) = \sup \left\{ \left\| \frac{x+\eta y}{2} \right\| + \left\| \frac{x-\eta y}{2} \right\| - 1 \quad ; \quad \| x \| \le 1 \, , \, \| y \| \le 1 \right\}$$

for any $\eta > 0$. X is said to be *uniformly smooth (U.S.)* if

$$\lim_{n \to 0} \frac{\rho_X(\eta)}{\eta} = 0$$

Theorem: *For any Banach space X, we have*

1. $\rho_{X^*}(\eta) = \sup\{ \eta\, \varepsilon / 2 - \delta_x(\varepsilon) \, ; \ 0 \le \varepsilon \le 2 \}$, *for any* $\eta > 0$;
2. $\lim\limits_{\eta \to 0} \dfrac{\rho_{X^*}(\eta)}{\eta} = \dfrac{\varepsilon_0(X)}{2}$
3. *X is uniformly convex if and only if X* is uniformly smooth.*

Note that one may reverse the roles of X and X^* to obtain for example

$$\rho'_X(0) = \lim_{n \to 0} \frac{\rho_X(\eta)}{\eta} = \frac{\varepsilon_0(X^*)}{2}$$

Using James non-squareness result, we see that X is superreflexive if and only if $\rho'_X(0) < \frac{1}{2}$

In 1978-79 Baillon proved that any uniformly smooth space X has the fixed point property. In fact, he precisely proved this conclusion provided $\rho'_X(0) < \frac{1}{2}$

In 1982, Turett published a paper where he proved that the condition $\rho'_X(0) < \frac{1}{2}$ implies the normal structure property. Unfortunately, Turett's proof is an exact reproduction of Baillon's original proof. Later on, this conclusion was strengthened by Prus (1988) and Khamsi (1987) to show that the condition $\rho'_X(0) < \frac{1}{2}$ implies the supernormal structure property (and hence the uniform normal structure property) for both X and its dual X^*. The proof is based on nonstandard techniques.

3.3.6 Karlovitz-Goebel Lemma

As mentioned before, if K is a minimal convex associated to a nonexpansive mapping, then little is known about its properties except the two properties discovered by Kirk. In 1975, independently Karlovitz and Goebel proved the following:

Theorem : *Let K be a subset of a Banach space X which is minimal with respect to being non empty, weakly compact, convex and T-invariant for some nonexpansive mapping T; and suppose $\{x_n\} \subset K$ is an a.f.p.s., i.e.*

$$\lim_{n\to\infty} \| x_n - T(x_n) \| = 0.$$

Then, for each $x \in K$,

$$\lim_{n\to\infty} \| x - x_n \| = diam(K).$$

Recall again that it was this property that allowed Karlovitz to prove that the space $X_{\sqrt{2}}$ has the fixed point property. In this theorem, *a.f.p.s.* are important. It is this that pushed Maurey to use ultrapowers.

From 1965 to 1980, most of the classical Banach spaces were investigated and proved to have the fixed point property. Missing to the call was c_0 and (general) lattice Banach spaces. During that period of time, it was strongly believed that any weakly compact convex subset of any Banach space have the fixed point property. It was Alspach (1980-81) who disproved this claim by constructing an isometry on a weakly compact convex subset of $L^1[0, 1]$ without a fixed point. This example set the stage to the second revolution in the theory (after the first one in 1965). Indeed, almost right after Alspach's example was made public, Maurey (1980) published his famous results (using nonstandard techniques).

The Open Problem *: It is still unknown whether reflexive (or superreflexive) Banach spaces have the fixed point property.*

3.4 Metric Fixed Point Theory in Metric Spaces

3.4.1 The Metric Convexity

The Banach Contraction Principle theorem is a metric result and does not depend on any linear structure. But, Kirk's fixed point theorem is strongly connected to the linear convexity structure of linear spaces. As early as 1965, many have tried to weaken this tiding. Takahashi (1970) was may be the first one to give a metric analogue to Kirk's theorem. His approach was based on defining a convexity in metric spaces extremely similar to the linear convexity.

Definition *: Let (M, d) be a metric space with $I = [0, 1]$. A mapping $W : M \times M \times I \to M$ is said to be a convex structure on M if for each $(x, y, \lambda) \in M \times M \times I$ and $z \in M$, we have*

$$d(z, W(x, y, \lambda) \leq \lambda d(z, x) + (1 - \lambda) d(z, y)$$

Sometimes the point $W(x, y, \lambda)$ is denoted by $\lambda x \oplus (1 - \lambda)y$ whenever the choice of the convexity mapping W is irrelevant. Using the convexity structure W, one will easily define a convex subset of M and prove similar properties of convex sets in the linear case. It is not hard to check that balls are convex sets. A very nice example of such situation is given by:

Let B be the open unit ball of the infinite Hilbert space H. On B, we consider the Poincare hyperbolic metric ρ:

$$\rho(x, y) = \inf_{\gamma} \int_0^1 \alpha(\gamma(t))\,_{\gamma}\gamma'(t))dt$$

where the infimum is taken over all piece-wise differentiable curves such that

$$\gamma(0) = x \text{ and } \gamma(1) = y$$

and where

$$\alpha(x, v) = \sup_{f \in \mathcal{F}} \| Df(x)(v) \|$$

with $\mathcal{F} = \{f : B \to B;\text{ holomorphic}\}$ (B, ρ) is a complete metric space (unbounded). Using the Mobius transformations, one can prove that

$$\rho(x, y) = \text{Argth}\,(1 - \sigma(x, y))^{1/2}$$

with
$$\sigma(x, y) = \frac{(1 - |x|^2 (1 - |y|^2)}{|1 - \langle x, y \rangle|^2}$$

Using some complicated computations, one may prove that for any x, y in B and $\lambda \in [0, 1]$, there exists a unique $z \in B$ such that

$$\rho(z,w) \leq \lambda \rho(x,w) + (1 - \lambda)\,\rho(y,w)$$

for any $w \in B$. In other words, we have $z = \lambda x \oplus (1 - \lambda)y$. In fact, the metric space (B, ρ) enjoys some geometric properties similar to uniformly convex Banach spaces. For example, (B, ρ) has a kind of weak compactness, that is the family of convex subsets of B (in the abstract sense) has the finite infinite intersection property. Moreover, we have

$$\begin{cases} \rho(a,x) \le r \\ \rho(a,y) \le r \\ \rho(x,y) \ge r\varepsilon \end{cases} \Rightarrow \rho\left(a, \frac{1}{2}x \oplus \frac{1}{2}y\right) \le r(1-\delta(r,\varepsilon))$$

for any *a, x, y* in *B* and any positive *r* and ε where

$$\delta(r,\varepsilon) = 1 - \frac{1}{r} Argth\left(\frac{\sinh(r(1+\frac{\varepsilon}{2}))\sinh(r(1-\frac{\varepsilon}{2}))}{\cosh(r)}\right)^{1/2}$$

It is easy to check that for $r > 0$ and $\varepsilon > 0$, we have $\delta(r,\varepsilon) > 0$. This is the analogue of the uniform convexity in the nonlinear case. It is not hard to check that convex subsets enjoy the normal structure property and that the Chebishev center is reduced to one point. Note that holomorphic mappings are nonexpansive mappings for the Poincare distance ½.

3.4.2 The Convexity Structures

In 1977, Penot was successful in giving an abstract formulation of Kirk's theorem via the *Convexity Structures*:

Definition: *Let M be an abstract set.A family* $\sum$ *of subsets of M is called a convexity structure if*

(i) the empty set $\phi \in \sum$
(ii) $M \in \sum$
(iii) $\sum$ *is closed under arbitrary intersections.*

The convex subsets of *M* are the elements of $\sum$. If *M* is a metric space, we will always assume that closed balls are convex. The smallest convexity structure which contains the closed balls is $\mathcal{A}(M)$ the family of admissible subsets of*M*.Recall that *A* is an admissible subset of *M* if it is an intersection of closed balls. Since Kirk's theorem involves some kind of compactness and the normal structure property, it was of no surprise that the generalized attempts did define these two concepts in metric spaces. Takahashi in his attempt considered compact metric spaces which was very restrictive. Penot on the other hand, defined compactness for convexity structures which leads to weak-compactness in the linear case. Indeed, a convexity structure $\sum$ is said to be *compact* if and only if every family of subsets of $\sum$ which has the finite intersection property has a nonempty intersection, i.e. if $(A_i)_{i\in I}$ with $Ai \in\sum$, then

$$\bigcap_{i\in I} A_i \neq \phi$$

provided $\bigcap_{i\in I_f} A_i \neq \phi$ for any finite subset I_f of I. Since the normal structure property is a metric notion, then it was not that difficult to extend it to convexity structures. Indeed, the convexity structure $\sum$ is said to be *normal* if and only if for any nonempty and bounded $A \in \sum$ not re-duced to one point, there exists $a \in A$ such that

$$\sup\{d(a, x)\ ;\ x \in A\} < \sup\{d(x, y)\ ;\ x, y \in A\}= diam(A)$$

Penot's formulation becomes

Theorem: *If (M, d) is a nonempty bounded metric space which possesses a convexity structure which is compact and normal, then every nonexpansive mapping $T : M \to M$ has a fixed point.*

Remark: *In the general Kirk's theorem, the Banach space is supposed to have normal structure which means that the family of all convex sets is normal. But, this family is a large one and contains the admissible sets. For example, the Banach space l^∞ is a wonderful example which illustrates the power behind Penot's formulation. Indeed, l^∞ fails to satisfy the normal structure property but $A(l^\infty)$ is compact and normal which implies the following theorem discovered separately by Sine and Soardi in 1979.*

Theorem : *If A is a nonempty admissible subset of l^∞, then every nonexpansive mapping $T : A \to A$ has a fixed point.*

Remark : *In the original proof of Kirk's theorem, the weak compactness is used to prove the existence of a minimal invariant* *set via Zorn's lemma. Gillespie and Williams (1979) showed that a constructive proof may be found which uses only countable compactness. In other words, the convexity structure $\sum$ is assumed to satisfy a countable intersection property, i.e. for any $(A_n)_{1\leq n}$, with $A_n \in \sum$, then $\bigcap_{n=1}^{\infty} A_n \neq \phi$ provided $\bigcap_{n=1}^{m} A_n \neq \phi$ for any $m \geq 1$. This weakening is very important since in many practical cases, we do not have a compactness generated by a topology but a compactness defined sequentially .The latest usually generates some kind of countable compactness. Note also that if the convexity structure $\sum$ is uniformly normal then it is countably compact. This is an amazing metric translation of a well known similar result in Banach spaces due to Maluta. It is natural to ask whether a convexity structure which is countably compact is basically compact. The answer is yes if we are dealing with uniform normal structure. In fact, if $\mathcal{A}(M)$ is countably compact and normal, then $\mathcal{A}(M)$ is compact.*

3.4.3 Generalized Metric Spaces

In the next example, we discuss another amazing translation of Kirk's theorem. Indeed, traditionally discrete techniques are used in metric spaces in our day-to-day research.The other way is not that common. But, let us generalize the common distance into the discrete case. For that let M be an arbitrary set and $\boldsymbol{V}$ a set with a binary operation which will be denoted by $\oplus$. We assume that $\oplus$ enjoys most of the properties that the classical addition does. In particular, we have a zero element $0 \in \boldsymbol{V}$ which satisfies $u \oplus 0 = 0 \oplus u = u$ for any $u \in \boldsymbol{V}$. We also assume that $\boldsymbol{V}$ is orered by $\leq$ such that $0 \leq u$ and $u \oplus v \leq u' \oplus v'$ whenever $u \leq u'$ and $v \leq v'$. Note that $\boldsymbol{V}$ is not totally behaving like the set of positive numbers since we do not have a multiplication operation on $\boldsymbol{V}$.

Definition : *Let $\boldsymbol{V}$ be a set as described above and M be an arbitrary set. The mapping $d : M \times M \to \boldsymbol{V}$ is called a generalized distance if*

(i) $d(x, y) = 0$ iff $x = y$

(ii) $d(x, y) = \tau (d(y, x))$, where τ is an involution, that is

$\tau(\tau(x)) = x$
(iii) $d(x, y) \leq d(x, z) \oplus d(z, y)$

To appreciate the power behind the above new concept, let us show how any ordered set is a generalized metric space. Let $(\mathcal{M}, \prec)$ be an ordered set. Set $\mathcal{V} = \{0, \alpha, \beta, 1\}$ ordered as : $0 \leq \alpha \leq 1$,
$0 \leq \beta \leq 1$ and α and β are not comparable. Define $\oplus$ as the max operation, i.e. $u \oplus v = \max(u, v)$. *Define the generalized distance d by*

(i) $d(x, x) = 0$
(ii) $d(x, y) = \alpha$, if $x \prec y$
(iii) $d(x, y) = \beta$, if $y \prec x$
(iv) $d(x, y) = 1$, if x and y are not comparable
(v) $\tau(0) = 0$, $\tau(\alpha) = \beta$, $\tau(\beta) = \alpha$ and $\tau(1) = 1$.

With $(d, \mathcal{M})$ becomes a generalized metric space. Armed with this distance, we can talk about balls (left and right to be precise) and admissible sets. It is quite amazing to show that if $(\mathcal{M}, \prec)$ is a complete lattice, then $\mathcal{A}(\mathcal{M})$ is compact and normal. Also, it is an amazing fact to see that non expansive maps are exactly the monotone increasing maps. The famous Tarski theorem becomes Kirk theorem. More can be said and little is known.

3.4.4 Structure of the Fixed Point Set

In the previous discussion, we looked at the fundamental problem of whether a fixed point exists. But once this problem is resolved, it is quite natural to wonder about the structure of the fixed point set. In particular, if we want to prove the existence of a common fixed point for two commuting mappings, we will need to find out more about the fixed point set. In particular, whether the main fixed point existence result is valid for such sets. Properties like retractions help solve this question. May be the first interesting result in this direction goes back to Bruck in the linear case. Inspired by hyperconvexity, we introduce the following

Definition: *Let M be a metric space. A subset A of M is said to be 1-local retract of M if for any family* $(B_i)_{i\in I}$ *of closed balls centered in A for which*

$$\bigcap_{i\in I} B_i \neq \phi$$

it is the case that

$$A\cap\left(\bigcap_{i\in I} B_i\right) \neq \phi$$

One of the most fundamental properties of 1-local retracts is

Proposition: *Let M be a metric space and suppose that* $\mathcal{A}(M)$ *is compact and normal. If N is a nonempty 1-local retract subset of M, then* $\mathcal{A}(N)$ *is compact and normal.*

The proof is technical but follows the same ideas used in the hyperconvex case. In order to connect this concept with our previous discussion, we have

Theorem: *Let M be a metric space and suppose that* $\mathcal{A}(M)$ *is compact and normal. If* $T : M \to M$ *is nonexpansive, then the fixed point set Fix(T) of T is a nonempty 1-local retract subset of M.*

This conclusion may be easily extended to finite commu- tative families.

Theorem : *Let M be a metric space and suppose that* $\mathcal{A}(M)$ *is compact and normal. Then, every finite family* $\mathcal{F}$ *of commuting nonexpansive self-mappings of M has a nonempty common fixed point set Fix(F).Moreover,* $Fix(\mathcal{F})$ *is a nonempty 1-local retract subset of M.*

The conclusion of this theorem is still valid for infinite families but the proof is extremely difficult and not intuitive. Again, it uses ideas developed by Baillon in the hyperconvex case.

3.5 Applications

3.5.1 First Cournot Equilibrium :

Two firms compete in a market producing at output levels q_1 and q_2. Each firm responds to the other firm's production level when choosing its own level of output. Specifically (with a_1, a_2, b_1, b_2 all positive),

$$q_1 = r_1(q_2) = a_1 - b_1 q_2$$
$$q_2 = r_2(q_1) = a_2 - b_2 q_1$$

but $q_i = 0$ if the above expression for q_i is negative. The function $r_i : \boldsymbol{R}_+ \to \boldsymbol{R}_+$ is firm i 's *reaction function.* Define $\bar{r} : \boldsymbol{R}_+^2 \to \boldsymbol{R}_+^2$ by $\bar{r}\ (q_1, q_2) = (r_1(q_2), r_2(q_1))$. The function $\bar{r}$ is a contraction with respect to the city-block metric if $b_1, b_2 < 1$.

$$\begin{aligned}
d(\bar{r}(q), \bar{r}(q')) &= |\bar{r}_1(q) - \bar{r}_1(q')| + |\bar{r}_2(q) - \bar{r}_2(q')| \\
&= |(a_1 - b_1 q_2) - (a_1 - b_1 q'_2)| + |(a_2 - b_2 q_1) - (a_2 - b_2 q'_1)| \\
&= b_1 |q'_2 - q_2| + b_2 |q'_1 - q_1| \\
&\leq \max\{b_1, b_2\} (|q_1 - q'_1| + |q_2 - q'_2|) \\
&= \max\{b_1, b_2\}\, d(q, q')
\end{aligned}$$

Therefore, we have an "existence and uniqueness result" for Cournot equilibrium : $\bar{r}$ has a unique fixed point q^* - a unique Cournot equilibrium - if each $b_i < 1$.

There are several things to be noted. First, note that while the condition $b_1, b_2 < 1$ is sufficient to guarantee the existence of an equilibrium, it is not necessary. Second, note that we could have obtained the same result and actually calculated the equilibrium production levels, by simply solving the two "response" equations simul- taneously. The condition $b_1, b_2 < 1$ is easily seen to be sufficient (and again, not necessary) to guarantee that the two-equation system has a solution. (Note that $b_1 b_2 \neq 1$ is in fact sufficient.) However, things might not be so simple if the response

functions are not linear or if there are more firms (and therefore more equations). We'll consider a second, nonlinear application shortly and we'll use the Method of Successive Approximations to compute the equilibrium.
A third thing to be noted is that we used the city-block metric instead of the Euclidean metric. This highlights an important and useful fact: a given function mapping a set X into itself may be a contraction according to one metric on X but not be a contraction according to other metrics. Recall that the definition of a fixed point and therefore whether a given point in the function's domain is a fixed point, does not depend on the metric we're using and in fact does not even require that X be endowed with any metric structure.

***The Method of Successive Approximations*:**

For any "initial" $x_0 \in X$, the sequence $x_n = Tx_{n-1}$ that we constructed in the proof of the Banach Fixed Point Theorem not only converges to the fixed point x^*, it converges to x^* monotonically: each successive term is closer to x^* than its predecessor was.

Consequently, starting from any arbitrary point in X, we can repeatedly apply the function T to the current "approximation" of x^*, obtaining a better approximation and the approximations converge to x^*. This often provides a straight forward method for computing x^*.

3.5.2 A surprising application: the PageRank algorithm :

Google's success as a search engine comes from its algorithm: the Page Rank algorithm. In this algorithm, one computes a fixed point of a linear operator on $\boldsymbol{R}^n$ which is a contraction and this fixed point (which is a vector) yields the ordering of the pages. In practice, the fixed point (which is an eigenvector of the eigenvalue 1) is calculated approximately as P_n for n sufficiently large. We invite the interested reader to look at the details in the vignette on *How Google works?*.

3.5.3 Application to image compression

The best way to store an image in memory is to store the color of each pixel. There are two problems with this method:

• It requires an enormous quantity of memory.

• If we try to enlarge the image, for instance for using it in a large poster, then the pixels will become larger squares and we will be missing information on how to fill the details in these squares.

What is the principle of image compression? It is to encode less information than the original image, but to do it in a clever way so that the eye cannot see that the image observed is deteriorated. Internet has increased the need for good image compression. Indeed, images slow down navigation on the web quite significantly. So, for internet navigation, it is good to have images encoded in files as small as possible. When you look at the image on your computer screen you cannot see that it has been deteriorated. But, if you try to enlarge it or to use it for a poster, then you immediately see that the quality is poor. There are several principles of image compression and one of the most familiar is *JPEG* which has become the standard for digital photos. The encoding of an image in *JPEG* format is also a mathematical algorithm.

In this vignette, we will concentrate on another method which has remained more experimental. This method, introduced by Barnsley, has been called iterated function systems. The underlying idea of the method is to approximate an image by geometric objects. To have sufficiently many objects available,we will not limit ourselves selves to the usual geometric objects which are lines and smooth curves but allow complicated fractal objects like the fern and the Sierpinski carpet.

(a) The fern (b) The Sierpiński carpet (c) A tree

Figure 1 : Three fractals

We will explain the idea of the compression process on the Sierpi´nski carpet of *Figure 1(b).* It looks a priori a complicated object. How to store it in computer memory in an economical way? The best is to store a program to reconstruct it when needed. And to build this program, we need to understand what characterizes this geometric object. Let us observe our Sierpi´nski carpet: it is the union of three Sierpinski carpets (i.e. 3 copies of itself), that are half its size (in width and height).Indeed, starting from the Sierpinski carpet, we can construct a second one with the follow ing procedure:

- We shrink the Sierpi´nski carpet to its half size from the lower left vertex.
- We make a second copy of this half Sierpi´nski carpet and glue it on the right.
- We make a third copy of this half Sierpi´nski carpet and glue it on the top.

The second figure we have built is identical to our initial Sierpinski carpet. So, the Sierpinski car pet is the fixed point for this process.

Let us now put that in mathematical terms. Note that the length of the base of the Sierpi´nski carpet in *Figure 1(b)* is equal to its height. Hence, we can take axes with the origin at the lower left corner of the Sierpi´nski carpet and units so that the base and height have both length 1. We also take the following affine transformations defined on $\boldsymbol{R}^2$:

$$T_1(x, y) = \left(\frac{x}{2}, \frac{y}{2}\right)$$

$$T_2(x, y) = \left(\frac{x}{2} + \frac{1}{2}, \frac{y}{2}\right)$$

$$T_3(x, y) = \left(\frac{x}{2} + \frac{1}{4}, \frac{y}{2} + \frac{1}{2}\right)$$

If S is the Sierpinski carpet, then we have

$$S = T_1(S) \cup T_2(S) \cup T_3(S).$$

Are there other subsets B of the plane which have the same property, namely

$$B = T_1(B) \cup T_2(B) \cup T_3(B) \quad \ldots\ldots\ldots\ldots\ldots(1)$$

We will experiment and see that there are none! So, we have characterized our Sierpinski carpet as the only subset B of the plane such that (1) is satisfied. What have we done? We have constructed a function which to a subset B of the plane associates the subset $T_1(B) \cup T_2(B) \cup T_3(B)$. Let us call this function W. It is defined by

$$B \mapsto W(B) = T_1(B) \cup T_2(B) \cup T_3(B) \ldots\ldots (2)$$

and we have observed that $S = W(S)$, *i.e.* S is a fixed point of this function.

We have announced that we would experiment that the Sierpinski carpet is the only fixed point of this function. Let us try with a square C_0 as in *Figure 2(a).* Its image is C_1 in *Figure 2(b).*

We apply the same process to C_1 and get C_2, C_3 - C_5. (*Figure 2(c)-(f)*). We observe three things:
(i) None of the sets $C_0, \ldots, C_5$ is a fixed point of W.
(ii) We could have continued the process indefinitely, thus, producing an infinite sequence of sets $\{C_n\}$, where $C_{n+1} = W(C_n)$.
(iii) The sequence $\{C_n\}$ seems to converge rapidly to the Sierpinski carpet. Indeed, our eye cannot distinguish C_{10} from S. So, instead of S on our image, the program reconstructing our image can just produce C_{10}. And, if a better resolution were needed, then we would use the same program and ask it to stop at C_{20} or C_{30}. So, the same very small program can reconstruct S to any precision.

Not only that, but we can experiment that it works with any initial set. A second example iterating a pentagon appears in *Figure3*. The same remarks *(i), (ii)* and *(iii)* as above apply to this example.

We have seen that the Banach fixed point theorem applies to contractions on complete metric spaces. We have defined the function W in (2) on subsets of the place. For the metric space, we will take K as the set of (closed) bounded subsets of the place. We will introduce a distance on K called the Hausdorff distance. The definition of the Hausdorff

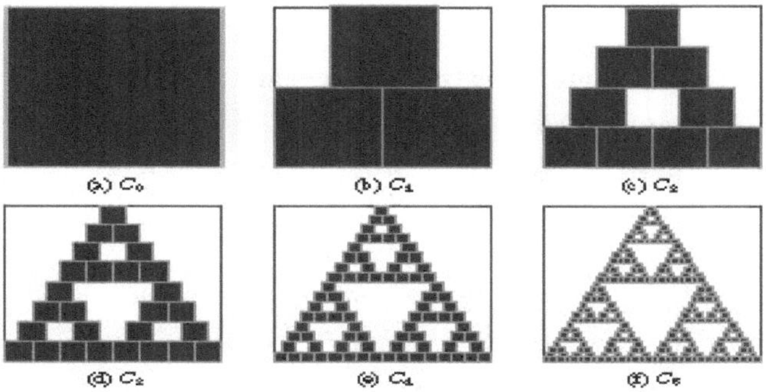

Figure 2: C_{O} and the first five iterations $C_1 – C_5$

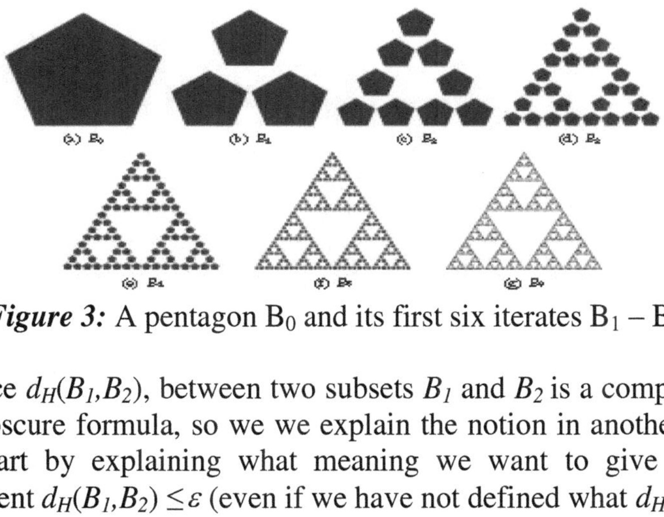

Figure 3: A pentagon B_0 and its first six iterates $B_1 – B_6$

distance $d_H(B_1,B_2)$, between two subsets B_1 and B_2 is a complicated and obscure formula, so we we explain the notion in another way. We start by explaining what meaning we want to give to the statement $d_H(B_1,B_2) \leq \varepsilon$ (even if we have not defined what $d_H(B_1,B_2)$ is). It just means that if our eye has a precision of ε , then it cannot distinguish B_1 from B_2. In mathematical terms, then $d_H(B_1,B_2) \leq \varepsilon$ should mean

$$\forall \ P \in B_1 \ \exists Q \in B_2 \ d(P,Q) \leq \varepsilon$$

and $\forall \; P' \in B_2 \; \exists Q' \in B_1 \; d(P',Q') \leq \varepsilon$ . ...(3)

(Here d is the usual Euclidean distance on $\boldsymbol{R}^2$.) This allows to give the indirect definition.

Definition : *The Hausdorff distance between two closed bounded sets B_1 and B_2 is the minimum of all $\varepsilon \geq 0$ such that (3) is satisfied.*

We can now convince ourselves that the function W is a contraction. Indeed, suppose that $d_H(B_1,B_2) \leq \varepsilon$. Then, we can show that $d_H(W(B_1), W(B_2)) \leq \frac{\varepsilon}{2}$. Let $P \in W(B_1)$. Then, there exist $i \in \{1, 2, 3\}$ and $P_1 \in B_1$ such that $P = T_i(P_1)$. Since $d_H(B_1, B_2) \leq \varepsilon$, there exists $Q_1 \in B_2$ such that $d(P_1,Q_1) < \varepsilon$. Let $Q = T_i(Q_1)$. Then $Q \in W(B_2)$ and $d(P,Q) = d(T_i(P_1), T_i(Q_1)) = \frac{1}{2}\,\mathrm{d}(P_1,Q_1) \leq \frac{\varepsilon}{2}$. Similarly, if we start with $P' \in W(B_2)$ thenthere exists $Q' \in W(B_1)$ such that $d(P',Q') \leq \frac{\varepsilon}{2}$.

This process has been adapted to the compression of real images. The method produces high quality images when the image has some fractal character. However, the rate of compression is not as good and not as flexible as the *JPEG* format. Also, the encoding process (transforming the image in a program to reconstruct it) is still too tedious to be of practical interest. However, the simplicity of its idea together with its power remains quite striking.

3.5.4 The implicit function theorem

A. Frechet differentiability : *Let X, Y be (real or complex) Banach spaces, $U \subset X$, U open, $x_0 \in U$, and $f : U \to Y$.*

Definition : *f is Frechet differentiable at x_0 is there exists $T \in L(X, Y)$ and $\sigma : X \to Y$ with*

$$\frac{\|\sigma(x)\|_Y}{\|x\|_X} \to 0 \quad \textit{uniformly as} \quad \|x\|_X \to 0$$

such that

$$f(x) - f(x_0) = T(x - x_0) + \sigma\ (x - x_0), \quad \textit{for all } x \in U.$$

The operator T is called the *Frechet derivative* of f at x_0, and is denoted by $f'(x_0)$. The function f is said to be Frechet differentiable in U if it is Frechet differentiable at every $x_0 \in U$.

It is straightforward to verify the Fr´echet derivative at one point, if it exists, is unique.

Lemma B : *Let X, Y be Banach spaces, let* $f : B_X(0, r) \to Y$ *be Frechet differentiable and* $\| f'(x) \|_{L(X,Y)} \leq \lambda$ *for every* $x \in B_X(0, r)$ *and some* $\lambda \geq 0$*. Then, f is Lipschitz continuous with Lipschitz constant less than or equal to* λ*.*

Lemma C : *Given two Banach spaces X and Y , the vector space X ×Y is a Banach space with any of the (equivalent) euclidean norms*

$$\| (x,y) \|_p = \left(\|x\|_X^p + \|y\|_Y^p\right)^{1/p},$$
$$\| (x,y) \|_\infty = \max\{ \| x \|_X ,\ \| y \|_Y \}, \quad (p \geq 1)$$

In the sequel, we will always use the ∞-norm, so that

$$B_{X\times Y}((x_0, y_0), r) = B_X(x_0, r) \times B_Y(y_0, r).$$

For X, Y, Z Banach spaces, given $T \in L(X,Z)$ and $S \in L(Y,Z)$, the operator $R : X \times Y \to Z$ defined by

$$R(x, y) = Tx + Sy$$

belongs to $L(X\times Y, Z)$. Conversely, any $R \in L(X \times Y, Z)$, has the above representation with $Tx = R(x, 0)$ and $Sy = R(0, y)$. It is then immediate to see that $L(X, Z) \times L(Y, Z)$ and $L(X \times Y, Z)$ are isomorphic Banach spaces. Given then $f : U \subset X \times Y \to Z$, U open, f Frechet differentiable at $u_0 = (x_0, y_0) \in U$, one easily checks that the partial derivatives $D_xf(u_0)$ and $D_yf(u_0)$ exist (that is, the Frechet derivatives of $f(\cdot, y_0) : X \to Z$ in x_0 and of $f(x_0, \cdot) : Y \to Z$ in y_0, respectively) and

$$f'(u_0)(x, y) = D_xf(u_0)(x) + D_yf(u_0)(y).$$

***Theorem D* [*Dini*] :** *Let X, Y, Z be Banach spaces, $U \subset X \times Y$ be an open set, $u_0 = (x_0, y_0) \in U$ and $F : U \to Z$. Assume that*

(a) F is continuous and $F(u_0) = 0$;
(b) $D_yF(u)$ exists for every $u = (x, y) \in U$;
(c) D_yF is continuous at u_0 and $D_yF(u_0)$ is invertible.

Then, there exists α, $\beta > 0$ for which $\overline{B}_X(x_0, \alpha) \times B_Y(y_0, \beta) \subset U$ and a unique continuous function $f : \overline{B}_X(x_0, \alpha) \to \overline{B}_Y(y_0, \beta)$ such that the relation

$$F(x, y) = 0 \quad \Leftrightarrow \quad y = f(x)$$

holds for all $(x, y) \in \overline{B}_X(x_0, \alpha) \times B_Y(y_0, \beta)$.

***Proof* :** Without loss of generality, we assume $x_0 = 0$ and $y_0 = 0$. Define

$$\Phi(x, y) = y - [D_yF(0, 0)]^{-1}F(x, y), \qquad (x, y) \in U.$$

By *(a)* , Φ is continuous from U into Y .

Since $[D_y \Phi(0, 0)]^{-1} (D_y \Phi(0, 0) - D_y \Phi(x, y))$,

by *(c)* there is $\gamma > 0$ small enough such that

$$||D_y \Phi(x, y)||_{L(Y)} \le \frac{1}{2},$$

for all $(x, y) \in B_X(0, \gamma) \times B_Y(0, \gamma) \subset U$.
Thus, *Lemma B* and the continuity of Φ entail the inequality

$$\| \Phi(x, y_1) - \Phi(x, y_2)\|_Y \leq \frac{1}{2} \|y_1 - y_2\|_Y ,$$
$$\|x\|_X, \|y_1\|_Y, \|y_2\|_Y \leq \beta < \gamma$$

Using now *(a)*, we find $0 < \alpha < \beta$ such that

$$\| \Phi(x, 0)\|_Y \leq \frac{\beta}{2} , \qquad \|x\|_X \leq \alpha .$$

Then, for $\|x\|_X \leq \alpha$ *and* $\|y\|_Y \leq \beta$,

$$\| \Phi(x, y)\|_Y \leq \| \Phi(x, 0)\|_Y + \| \Phi(x, y) - \Phi(x, 0)\|_Y$$
$$\leq \frac{1}{2} (\beta + \|y\|_Y) \leq \beta$$

Therefore, the continuous map $\Phi: \overline{B}_X(0, \alpha) \times \overline{B}_Y (0, \beta) \rightarrow \overline{B}_Y (0, \beta)$ is a contraction on $\overline{B}_Y (0, \beta)$ uniformly in $\overline{B}_X(0, \alpha)$. By *corollary D* (page 78), there exists a unique continuous function $f : \overline{B}_X(0, \alpha) \rightarrow \overline{B}_Y (0, \beta)$ such that $\Phi(x, f(x)) = f(x)$, that is, $F(x, f(x)) = 0$

Obviously, the thesis still holds replacing in the hypotheses closed balls with open balls.

Corollary: *Let the hypotheses of Theorem D hold. If in addition F is Frechet differentiable at $u_0 = (x_0, y_0)$, then f is Fr´echet differentiable at x_0 and*

$$f'(x_0) = -[D_yF(u_0)]^{-1}D_xF(u_0).$$

E. Location of zeros : Let X, Y be Banach spaces, and $f : B_X(x_0, r) \rightarrow Y$ be a Frechet differentiable map. In order to find a zero for f, the idea is to apply an iterative method constructing a sequence x_n (starting from x_0) so that x_{n+1} is the zero of the tangent of f at x_n. Assuming that $f'(x)^{-1} \in L(Y,X)$ on $B_X(x_0, r)$, one has

$$x_{n+1} = x_n - f'(x_n)^{-1} f(x_n) \ldots\ldots\ldots\ldots(1)$$

provided $x_n \in B_X(x_0, r)$ for every n. This procedure is known as the *Newton method.* However, for practical purposes, it might be complicated to invert f' at each step. So, one can try the modification

$$x_{n+1} = x_n - f'(x_0)^{-1} f(x_n). \quad \ldots\ldots\ldots\ldots\ldots(2)$$

Clearly, using (2) in place of (1), a lower convergence rate is to be expected. The following result is based on (2).

Theorem : *Let X, Y be Banach spaces and $f: B_X(x_0, r) \to Y$ be a Frechet differentiable map. Assume tha, for some $\lambda > 0$,*

(a) $f'(x_0)$ is invertible;
(b) $\|f'(x) - f'(x_0)\|_{L(X,Y)} \le \lambda \|x - x_0\|_X$, for all $x \in B_X(x_0, r)$;
(c) $\mu := 4\lambda \|f'(x_0)^{-1}\|^2_{L(Y,X)} \|f(x_0)\|_Y \le 1$;
(d) $s := 2\|f'(x_0)^{-1}\|_{L(Y,X)} \|f(x_0)\|_Y < r$.

Then there exists a unique $\bar{x} \in \bar{B}_X(x_0, s)$ such that $f(\bar{x}) = 0$.

Proof : Define $\Phi: \bar{B}_X(x_0, s) \to X$ as $\Phi(x) = x - f'(x_0)^{-1} f(x)$. Then,

$$\begin{aligned}\|\Phi'(x)\|_{L(X)} &\le \|f'(x_0)^{-1}\|_{L(Y,X)} \|f'(x_0) - f'(x)\|_{L(X,Y)} \\ &\le \lambda s \|f'(x_0)^{-1}\|_{L(Y,X)} = \frac{\mu}{2}\end{aligned}$$

Hence, Φ is Lipschitz with Lipschitz constant less than or equal to $\mu/2 < 1/2$. Moreover,

$$\|\Phi(x_0) - x_0\|_X \le \|f'(x_0)^{-1}\|_{L(Y,X)} \|f(x_0)\|_Y = \frac{s}{2}$$

which in turn gives

$$\begin{aligned}\|\Phi(x) - x_0\|_X &\le \|\Phi(x) - \Phi(x_0)\|_X + \|\Phi(x_0) - x_0\|_X \\ &\le \frac{\mu}{2}\|x - x_0\|_X + \frac{s}{2} \le s.\end{aligned}$$

Hence, Φ is a contraction on $\overline{B}_X(x_0, s)$. By *Banach's contraction principle* there exists a unique $\bar{x} \in \overline{B}_X(x_0, s)$ such that $\Phi(\bar{x}) = \bar{x}$, which implies $f(\bar{x}) = 0$.

Concerning the convergence speed of x_n to $\bar{x}$, by virtue of the remark after *Banach's contraction principle* , we get

$$||x_n - \bar{x}||_X \leq \frac{s\mu^n}{(2-\mu)2^n}$$

Also, since

$$x_{n+1} - \bar{x} = f'(x_0) - (f'(x_0) - f'(x_n)(x_n - \bar{x}) + o(\|x_n - \bar{z}\ \|X)$$

it follows that

$$||x_{n+1} - \bar{x}||_X = \frac{\mu}{2}||x_n - \bar{x}\ ||_X + o(||x_n - \bar{z}\ ||_X)$$

Hence,

$$||x_{n+1} - \bar{x}||_X \leq c||x_n - \bar{x}\ ||_X$$

for some $c \in (0, 1)$. for all large n. This is usually referred to as *linear convergence* of the method.

3.5.5 Ordinary differential equations in Banach spaces

A . The Riemann integral : Let X be a Banach space, $I = [\alpha, \beta] \subset$ ***R***. The notion of Riemann integral and the related properties can be extended with no differences from the case of real-valued functions to X-valued functions on I. In particular, if $f \in C(I,X)$ then f is Riemann integrable on I,

$$\left\| \int_\alpha^\beta f(t)dt \right\|_X \leq \int_\alpha^\beta \| f(t) \|_X \, dt$$

and

$$\frac{d}{dt}\int_{\alpha}^{t} f(y)dy = f(t), \text{ for all } t \in I$$

Recall that a function $h : I \to X$ is differentiable at $t_0 \in I$ if the limit

$$\lim_{t \to t_0} \frac{h(t) - h(t_0)}{t - t_0}$$

exists in X. This limit is the derivative of h at t_0 and is denoted by $h'(t_0)$ or $\frac{d}{dt}h(t_0)$. If $t_0 \in (\alpha, \beta)$ we recover the definition of *Frechet differentiability*.

It is easy to see that if $h'(t) = 0$ for all $t \in [\alpha, \beta]$, then $h(t)$ is constant on $[\alpha, \beta]$. Indeed, for every $\Lambda \in X^*$, we have $(\Lambda o\ h)'(t) = \Lambda' h(t) = 0$, that implies $\Lambda(h(t) - h(\alpha)) = 0$, and from the *Hahn-Banach theorem* there is $\Lambda \in X^*$ such that $\Lambda(h(t) - h(\alpha)) = ||h(t) - h(\alpha)||$.

B. The Cauchy problem : Let X be a Banach space, $U \subset \boldsymbol{R} \times X$, U open, $u_0 = (t_0, x_0) \in U$, $f : U \to X$ continuous. The problem is to find a closed interval I with t_0 belonging to the interior of I and a differentiable function $x : I \to X$ such that

$$\begin{cases} x'(t) = f(t, x(t)), & t \in I \\ x(t_0) = x_0 \end{cases} \quad \text{.........(1)}$$

It is apparent that such x is automatically of class C^1 on I. Also, it is readily seen that (1) is equivalent to the integral equation

$$x(t) = x_0 + \int_{t_0}^{t} f(y, x(y))\, dy,\ t \in I. \quad \text{.........(2)}$$

Namely, x is a solution to (1) if and only if it is a solution to (2).

***Theorem C* [*Local solution*] :** *Assume the following hypotheses:*

(a) f is continuous;
(b) The inequality $\|f(t, x_1) - f(t, x_2)\|_X \leq k(t)\|x_1 - x_2\|_X$, for all $(t, x_1), (t, x2) \in U$ holds for some $k(t) \in [0, \infty]$;
(c) $k \in L^1((t_0 - a, t_0 + a))$ for some $a > 0$;
(d) There exist $m \geq 0$ and $\overline{B}_{\mathbf{R}\times X}(u_0, s) \subset U$ such that $\|f(t, x)\|_X \leq m$, $\forall\ (t, x) \in \overline{B}_{\mathbf{R}\times X}(u_0, s)$.

Then, there is $\tau_0 > 0$ such that for any $\tau < \tau_0$, there exists a unique solution $x \in C^1(I_\tau, X)$ to the Cauchy problem B, with $I_\tau = [t_0 - \tau, t_0 + \tau]$.

Remark : *Notice first that from (b), since U is open, k is defined in a neighborhood of zero. If k is constant then (c)-(d) are automatically satisfied. Indeed, for $(x, t) \in \overline{B}_{\mathbf{R}\times X}(u_0, s)$, we have*

$$\|f(t, x)\|_X \leq ks + \max_{|t-t_0|\leq s} \|f(t, x_0)\|_X$$

Also, (d) is always true if X is finite-dimensional, for closed balls are compact. In both cases, setting

$$s_0 = \sup\{\sigma > 0 ;\ \overline{B}_{\mathbf{R}\times X}(u_0, \sigma) \subset U\}$$

we can choose any $s < s_0$.

Proof : Let $r = \min\{a, s\}$ and set $\tau_0 = \min\{r, \frac{r}{m}\}$. Select then $\tau < \tau_0$ and consider the complete metric space $Z = \overline{B}_{C(I_\tau, X)}(x_0, r)$ with the metric induced by the norm of $C(I_\tau, X)$ (here x_0 is the constant function equal to x_0). Since $\tau < r$, if $z \in Z$ then $(t, z(t)) \in \overline{B}_{\mathbf{R}\times X(u_0, r)} \subset U$ for all $t \in I_\tau$. Hence, for $z \in Z$, define

$$F(z)(t) = x_0 + \int_{t_0}^{t} f(y, z(y))dy \ , \quad t \in I_\tau$$

Observe that

$$\begin{aligned}
&\sup_{t \in I_\tau} ||F(z)(t) - x_0||_X \\
&\leq \sup_{t \in I_\tau} \left| \int_{t_0}^{t} \| f(y, z(y)) \|_X \ dy \right| \\
&\leq m\tau \\
&\leq r
\end{aligned}$$

We conclude that F maps Z into Z. The last step is to show that F^n is a contraction on Z for some $n \in N$. By induction on n, we show that, for every $t \in I_\tau$,

$$||F^n(z_1)(t) - F^n(z_2)(t)||_X \leq \frac{1}{n!} \left| \int_{t_0}^{t} k(y)dy \right|^n || z_1 - z_2 ||_{C(I_\tau, X)} \cdots\cdots(3)$$

For $n = 1$, it holds easily. So, assume it is true for $n - 1, n \geq 2$. Then, taking $t > t_0$ (the argument for $t < t_0$ is analogous),

$$\begin{aligned}
&|| F^n(z_1)(t) - F^n(z_2)(t)||_X \\
&= ||F(F^{n-1}(z_1))(t) - F(F^{n-1}(z_2))(t)||_X \\
&\leq \int_{t_0}^{t} ||f(y, F^{n-1}(z_1)(y)) - f(y, F^{n-1}(z_2)(y))||_X dy \\
&\leq \int_{t_0}^{t} k(y)||F^{n-1}(z_1)(y) - F^{n-1}(z_2)(y)||_X dy \\
&\leq \left[\int_{t_0}^{t} k(y) \left(\int_{t_0}^{y} k(w) \right)^{n-1} dy \right] || z_1 - z_2 ||_{C(I_\tau, X)}
\end{aligned}$$

$$= \frac{1}{n!}\left(\int_{t_0}^{t} k(y)dy\right)^n \quad || z_1 - z_2 ||_{C(I_\tau, X)}$$

Therefore, from (3) we get

$$|| F^n(z_1) - F^n(z_2)||_{C(I_\tau, X)} \leq \frac{1}{n!} || k||^n_{L^1(I_\tau)} || z_1 - z_2 ||_{C(I_\tau, X)}$$

which shows that for n big enough F^n is a contraction. By *corollary E* (page 78), we conclude that F admits a unique fixed point which is clearly the (unique) solution to the integral equation (2) and hence, to (1).

3.5.6 Semilinear equations of evolution

A. Strongly continuous semigroups : *Let X be a Banach space. A one-parameter family S(t), (t ≥ 0) of bounded linear operators on X is said to be a strongly continuous semigroup(C_0 – semi-group, for short) if*

(a) S(0) = ***I*** *(identity operator on X);*
(b) S(t + s) = S(t)S(s) for every t, s ≥ 0;
(c) lim $_{t \to 0}$ *S(t)x = x for every x ∈ X (strong continuity).*

As a quite direct application of the uniform boundedness theorem, there exist $\omega \geq 0$ and $M \geq 1$ such that

$$||S(t)||_{L(X)} \leq Me^{\omega t}, \qquad \forall \quad t \geq 0.$$

This, in turn, entails the continuity of the map $t \mapsto S(t)x$ from $[0, \infty)$ to X, for every fixed $x \in X$.

The *linear operator* A of domain

$$\mathcal{D}(A) = \left\{ x \in X \; : \; \lim_{t \to 0} \frac{S(t)x - x}{t} \quad \text{exists} \right\}$$

defined by

$$Ax = \lim_{t\to 0}\frac{S(t)x - x}{t}, \quad \forall\ x \in\ = \mathcal{D}(A)$$

is the *infinitesimal generator* of the semigroup $S(t)$.

Proposition : *A is a closed linear operator with dense domain. For every fixed* $x \in \mathcal{D}(A)$, *the map* $t \mapsto S(t)x$ *belongs to* $C^1([0,\infty), \mathcal{D}(A))$ *and*

$$\frac{d}{dt}S(t)x = AS(t)x = S(t)Ax.$$

B. We consider the following semilinear Cauchy problem in X:

$$\begin{cases} x'(t) = Ax(t) + f(t,x(t)),\ 0 < t \le T \\ x(0) = x_0 \in X \end{cases} \qquad \ldots\ldots..(1)$$

where A is the infinitesimal generator of a strongly continuous semigroup $S(t)$ and $f : [0, T] \times X \to X$ is continuous and uniformly Lipschitz continuous on X with Lipschitz constant $\lambda \geq 0$.

Definition : *A function* $x : [0, T] \to X$ *is said to be a classical solution to (1) if it is differentiable on* $[0, T]$, $x(t) \in \mathcal{D}(A)$ *for every* $t \in [0, T]$ *and (1) is satisfied on* $[0, T]$.

If x is a classical solution, it is necessarily unique and it is given by

$$x(t) = S(t)x_0 + \int_0^t S(t-s)f(s, x(s))\, ds. \quad \ldots\ldots\ldots(2)$$

Definition : *A function* $x : [0, T] \to X$ *is said to be a mild solution to (1) if it continuous on* $[0, T]$ *and fulfills the integral equation (2).*

Theorem : *For every $x_0 \in X$ the Cauchy problem (1) has a unique mild solution. Moreover the map $x_0 \to x(t)$ is Lipschitz continuous from X into C([0, T],X).*

Proof : For a given $x_0 \in X$, define the map $F : C([0, T], X) \to C([0, T], X)$ by

$$F(x)(t) = S(t)x_0 + \int_0^t S(t-s)f(s, x(s))\, ds.$$

Then, we have

$$||F(x)(t) - F(y)(t)||_X$$

$$\leq \lambda M t\, ||x - y||_{C([0,T],X)}$$

where $M = sup_{t \in [0,T]} ||S(t)||_{L(X)}$.

By inductive argument, we get

$$||F^n(x)(t) - F^n(y)(t)||_X$$

$$\leq \frac{(\lambda MT)^n}{n!} ||x - y||_{C([0,T],X)}.$$

Hence, for $n \in N$ big enough F^n is a contraction. So, by *Corollary E* (page 77), F has a unique fixed point in $C([0, T], X)$ which is clearly the desired mild solution to the Cauchy problem (1).

To complete the proof, let y be the unique mild solution corresponding to the initial value y_0. Then,

$$||x(t) - y(t)||_X$$

$$\leq M||x_0 - y||_X + \lambda M \int_0^t ||x(s) - y(s)||_X \, ds.$$

By the *Gronwall Lemma*, we get at once

$$||x(t) - y(t)||_X \leq M\, e^{\lambda MT} ||x_0 - y_0||_X, \quad \forall \; t \in [0, T]$$

which entails the Lipschitz continuity of the map $x_0 \to x(t)$.

Chapter 4
Order-Theoretic Fixed Point Theory

4.1 Introduction

Fixed point theory serves as an essential tool for various branches of mathematical analysis and its applications. Loosely speaking, there are three main approaches in this Theory out of which the first two i.e., *the topological approach* in which one utilizes the topological properties of the underlying spaces and continuity of self-maps and *the metric approach* in which one makes use of the metric properties of the underlying spaces and self-maps have been discussed in chapter 2 and 3 respectively. Finally, the third approach is *the order-theoretic approach.* In this approach, one assumes that the underlying space is a certain type of a *poset (partially ordered set)* and seeks conditions under which an order preserving self-map has a fixed point. The main objective of this chapter is to sketch the elements of this approach and its applications.

We begin our exposition with an overview of various completeness conditions for posets; these play an essential role in the order-theoretic approach to fixed point theory. The first fixed point result that we report here is the *Tarski-Kantorovitch Theorem* - this is one of the few fixed point theorems that actually find fixed points by means of a constructive (iterative) process. We then move on to the most famous order-theoretic fixed point theorem, namely, *the Knaster-Tarski Theorem.* This result tells us not only that any order-preserving map on a complete lattice has a fixed point but also that the collection of all fixed points of such a map is a complete lattice itself (relative to the lattice order inherited from the domain of the self-map). These aim to demonstrate in what way one can make use of order-theoretic fixed point theory in practice.

4.2 Completeness Conditions for Posets

Order-theoretic fixed point theory is concerned with the existence (and properties) of fixed points of self-maps defined on certain types of ordered sets. To this end, one uses the structure of ordered sets (acting as both the domain and codomain) in tandem with that of the involved self-maps. And for this, we almost always need our ordered sets to possess some sort of a "completeness" property. Before we move on to order-theoretic fixed point theory proper, therefore, we shall consider in this section various types of completeness notions for posets.

We have so far considered mainly two notions of "completeness" for a poset, namely, chain-completeness and completeness as a lattice. Fixed point theory is actually very pleasant when one deals with self-maps defined on complete lattices. Unfortunately, in many applications we are forced to deal with posets that are not even lattices. Chain-completeness property is more commonly encountered in practice but there are nonetheless instances in which we need to deal with posets that are complete in an alternative sense. We begin this section by introducing two such alternatives to the chain-completeness property and then revisit this property providing this time a fulleraccount for it.

4.2.1 Conditionally Complete Posets

Recall that every nonempty subset of real numbers that is bounded from above has a supremum. (In fact, this property - often called the "completeness axiom" - singles $\boldsymbol{R}$ out among all ordered fields (up to ordered field isomorphism). It is, thus, natural to formulate this property in the general context of posets.

Definition : *Let (X, $\succeq$) be a poset. We say that (X, $\succeq$) is conditionally complete if $\bigvee S$ exists for every nonempty subset S of X such that $x \succeq S$ holds for at least one x in X.*

Put diffierently a poset (X, $\succeq$) is conditionally complete iff any nonempty subset S of X with an $\succeq$ - upper bound in X has a $\succeq$ - supremum in X. We note that that this notion is distinct from chain-completeness.

Proposition : *Let (X, $\succeq$) be a conditionally complete poset. Then, $\bigvee T$ exists for every non empty subset T of X such that $T \succeq x$ holds for some $x \in X$.*

Corollary : *If (X, $\succeq$) is a conditionally complete poset, then, so is (X, $\preceq$) .*

Proposition : *The product ($X \times Y$, $\succeq$) of two posets (X , $\succeq_X$) and (Y, $\succeq_Y$) is conditionally complete if and only if both (X , $\succeq_X$) and (Y , $\succeq_Y$) are conditionally com- plete.*

4.2.2 Countably Chain-Complete Posets

Chain-completeness is by far the most commonly used order-completeness property in fixed point theory. As we shall see, however, iterative fixed point theorems can make do with the weaker version of this property in which we posit the existence of suprema only for countable chains.

Definition : *Let (X, $\succeq$) be a poset. We say that (X, $\succeq$) is countably chain-complete if every countable $\succeq$- chain in X has a $\succeq$ - supremum in X.*

Before considering some examples, we will provide two characterizations of the countable chain-completeness property. To this end, we shall need the following bit of terminology from order theory.

Definition : *Let* $(X, \succsim)$ *be a preordered set. We say that a sequence* (x_m) *in* X *is* $\succsim$ *- increasing if* $x_{m+1} \succsim x_m$ *for each positive integer* m *and* $\succsim$ *- decreasing iff* $x_m \succsim x_{m+1}$ *for each* m.

Note: *Given a preordered set* $(X, \succsim)$, *a sequence* (x_m) *in* X *is* $\succsim$ *- increasing iff the map* $m \mapsto x_m$ *is an order-preserving map from* $(N, \geq)$ *into* $(X, \succsim)$.

Proposition : *The following are equivalent for any poset* $(X, \succeq)$ *:*

(a) $(X, \succeq)$ *is countably chain-complete;*
(b) $\bigvee\{x, x_2, \ldots\}$ *exists for any* $\succeq$ *- increasing sequence* (x_m) *in* X;
(c) $\bigvee D$ *exists for every countable* $\succeq$ *- directed subset* D *of* X.

This proposition characterizes countable chain-completeness by means of one seemingly weaker and one seemingly stronger property. In particular, the equivalence of *(a)* and *(b)* often simplifies the process of verifying the countable chain-completeness of a poset while that of *(b)* and *(c)* is useful for showing that a poset is not countably chain-complete.

4.2.3 Chain-Complete Posets, Revisited

Recall that a chain-complete poset is one in which every chain has a supremum. We know that every nonempty compact subset of $\boldsymbol{R}^n$ is chain-complete but we are then unable to extend this result outside the realm of Euclidean spaces. We are now in a position to remedy this situation.

Let us first introduce the following terminology.

Definition : *A triplet $(X, d, \succeq)$ is said to be a partially ordered (compact) metric space if (X, d) is a (compact) metric space and $(X, \succeq)$ is a poset such that $\succeq$ is closed- continuous (relative to the metric topology of (X, d)*

Proposition: *Let $(X, d, \succeq)$be a partially ordered compact metric space. Then, $(X, \succeq)$ is a chain-complete poset.*

Example : Let $(X, \geq)$ be a partially ordered linear space, that is, $\geq$ is a partial order on the linear space X which is defined as

$$x \geq y \quad \mathit{iff} \quad x \in y + C$$

for some convex cone in X with $C \cap -C = \{0\}$. If X is a Banach space (that is, it is a complete normed linear space), and if C is closed in X, then $(X, \geq)$ is said to be *an ordered Banach space*. As the closedness of C entails that $\geq$ is closed-continuous, an immediate application of the above proposition yields: *Every nonempty compact subset of an ordered Banach space is chain - complete* (relative to the vector partial order of the space).

Example : Consider the poset $(\boldsymbol{B}(X), \geq)$ we considered in the previous example. As $\boldsymbol{B}(X)$ is a Banach space, and it is easily checked that $\geq$ is a closed-continuous partial order on $\boldsymbol{B}(X)$, the poset $(B(X), \geq)$ is an ordered Banach space. This space is not chain-complete, while it is conditionally complete. However, by our previous example, we can say that $(\mathcal{F}, \geq)$ is chain complete for any nonempty compact subset $\mathcal{F}$ of $\boldsymbol{B}(X)$.

We now turn to a characterization of the chain-completeness property in terms of directed sets. This result, which is due to Markowsky (1976) is the exact analogue of the equivalence of conditions *(a)* and *(c)* in the proposition of page 131 in the absence of the countability hypothesis.

Proposition : *A poset $(X, \succeq)$ is chain-complete if and only if $\bigvee D$ exists for every $\succeq$ - directed subset D of X.*

Our final order of business in this section is to explore the connection between the chain-completeness of a lattice with its completness. It is plain that the latter implies the former. However, it turns out that a chain-complete lattice is in some sense half way being a complete lattice.

***Proposition** : Let $(X, \succeq)$ be a chain-complete lattice. Then, $\bigvee S$ exists in X for every nonempty subset S of X.*

***Corollary** : Let $(X, \succeq)$ be a lattice. Then, $(X, \succeq)$ is a complete lattice if and only if both $(X, \succeq)$ and $(X, \preceq)$ are chain-complete lattices.*

4.3 Iterative Fixed Point Theorems

4.3.1 A Prelude

Loosely put order-theoretic fixed point theory attempts to identify those situations in which an order-preserving self-map on a poset has a fixed point and for the cases in which this can be done tries to analyze the order structure of the set of all fixed points of the self-map at hand. To hint at what sort of things can be accomplished in this regard, we provide next three order-theoretic fixed point theorems. As they deal only with finite posets, these results are rather simplistic we shall obtain far reaching generalizations of them subsequently. Our aim here is just to illustrate the basic nature of order-theoretic arguments in fixed point theory.

***Proposition** : Let $(X, \succeq)$ be a finite poset and f a $\succeq$- preserving self - map on X. If there exists an x_* in X with $f(x_*) \succeq x_*$, then f has a fixed point. Furthermore, $Fix(f) \bigcap x_*^{\uparrow}$ has a $\succeq$ - minimum element.*

It is sometimes possible to extend an order-theoretic fixed point theorem in the case of a self-map that satisfy certain properties to the case of a finite collection of self-maps all of which satisfy that property. The key requirement for this is that these self-maps commute.

Definition: *Let X be a nonempty set and $\mathcal{F}$ a collection of self-maps on X. We say that $\mathcal{F}$ is a commutative collection if $\mathcal{F} \neq \phi$ and $f \circ g = g \circ f$ for every f and g in $\mathcal{F}$.*

The following is a typical example of how one can turn a fixed point theorem for a self-map into a common fixed point theorem for a commutative collection of self-maps.

Proposition : *Let $(X, \succeq)$ be a finite poset and $\mathcal{F}$ a commutative collection of $\succeq$ - preserving self-maps on X. If there exists an x_* in X with $f(x_*) \succeq x_*$ for each $f \in \mathcal{F}$, then $\mathcal{F}$ has a fixed point. Furthermore, Fix($\mathcal{F}$) $\bigcap x_*^{\uparrow}$ has a $\succeq$ - minimum element.*

Note: There is actually nothing too surprising about requiring the set $\mathcal{F}$ to be commutative in common fixed point theorems such as the above proposition . After all, if $x \in$ Fix($\mathcal{F}$) for some nonempty collection $\mathcal{F}$ of self-maps, then $f(g(x)) = f(x) = x = g(x) = g(f(x))$.That is, the members of $\mathcal{F}$ must commute on the set of fixed points of $\mathcal{F}$.

Proposition : *Let $(X, \succeq)$ be a finite poset and Γ a upper $\succeq$ - preserving self – correspondence on X. If there exists an x_* in X with $y \succeq x_*$ for some $y \in \Gamma(x_*)$ then Γ has a fixed point.*

4.3.2 The Tarski-Kantorovitch Fixed Point Theorem

As simple as it is, the basic argument behind the first proposition of page 126 is quite telling. We begin from a point x_* in the domain of f with $f(x_*) \succeq x_*$ and then observe that the iterations of x_* under f yields an $\succeq$ - increasing sequence, namely, $(f_m(x_*))$. Let $S = \{f(x_*), f^2(x_*), \ldots\}$ and notice that, only on the basis of the $\succeq$ - preserving nature of f we can conclude that

$$f(\vee S) \succeq \vee f(S) = \vee S \ldots\ldots\ldots\ldots(1)$$

The trouble is to show that $\vee S \succeq f(\vee S)$. When X is finite, this holds readily, because, then, S is finite and hence it has a $\succeq$ - maximum, that is, $\vee S \in S$. But, then $f(\vee S) \in S$ and hence , $\vee S \succeq (\vee S)$ holds. When X is not finite, of course, this argument breaks down. However, (1) shows that if we were to have $f(\vee S) = \vee f(S)$, then we would be guaranteed of a fixed point even if X is not finite (but has enough structure to warrantthat $\vee S$ exists). This observation prompts the notion of order-continuity.

Definition : *Let $(X, \succeq)$ be a countably chain-complete poset and f a self-map on X. We say that f is $\succeq$ - continuous if*

$$f(\vee\{x_1, x_2, \ldots\}) = \vee \{f(x_1), f(x_2), \ldots\}$$

for every $\succeq$ - increasing sequence (x_m) in X.

It is easy to see that $\succeq$ - continuity is a more stringent requirement than $\succeq$ - preservation. Indeed, where $(X, \succeq)$ is a poset, if f is a $\succeq$ - continuous self-map on X, then, for any x and y in X with $x \succeq y$, we have $x = x \vee y$ and hence,

$$f(x) = f(x \vee y) = f(x) \vee f(y)$$

because $(y, x, x, \ldots\ldots)$ is a $\succeq$ - preserving sequence in X and this implies $f(x) \succeq f(y)$.

Conclusion: *Every order-continuous self-map is order-preserving. The converse is, of course, false.*

Lemma : *Let $(X, \succeq)$ be a countably chain-complete poset and f a self-map on X. Then, f is $\succeq$ - continuous if and only if $f(\vee D) = \vee f(D)$ for every countable $\succeq$ - directed subset D of X.*

The Tarski-Kantorovitch Theorem *: Let $(X, \succeq)$ be a countably chain-complete poset and f a $\succeq$ - continuous self-map on X. If there exists an x_* in X with $f(x_*) \succeq x_*$, then ($Fix(f) \bigcap x_*^{\uparrow}$, $\succeq$) is a countably chain-complete poset with a $\succeq$ - minimum element.*

It is easy to extend the Tarski-Kantorovitch Theorem to the context of finite commutative collections of order-continuous self-maps. In our next result, we do this for countable commutative collections, thereby obtaining a substantial generalization of the Tarski-Kantorovitch Theorem.

Corollary *: Let $(X, \succeq)$ be a countably chain-complete poset and $\mathcal{F}$ a countable commutative collection of $\succeq$ - continuous self-maps on X. If there exists an x_* in X with $f(x_*) \succeq x_*$ for each $f \in F$, $(Fix(f) \bigcap x_*^{\uparrow}, \succeq)$ is a countably chain-complete poset with a $\succeq$ - minimum element.*

4.3.3 Applications

4.3.3 (a) Solving Functional Equations

Let T be an arbitrary nonempty set, h a self-map on T and H a bounded real map on $T \times R$: Consider the following problem:

Under what conditions on h and H is there a bounded real map g on T that satisfies the functional equation

$$g = H(\,.\,, g \circ h)$$

that is, $g(t) = H(t, g(h(t)))$ for every $t \in T$. Such problems arise frequently when working with differential and / or integral equations and in certain dynamic optimization problems in which a solution is characterized by means of a functional equation.

Put precisely, let T be an arbitrary nonempty set, h a self-map on T and $H : T \times R \to R$ a bounded function which is increasing and continuous in its second component. We would like to find out if there is a bounded real function g on T such that

$$g(t) = H(t, g(h(t))) \text{ , for every } \quad t \in T \text{.........(2)}$$

We can solve this problem by means of fixed point theory in the following manner.

Define

$$X = \{ g \in \boldsymbol{B}(T) : ||g||_{\infty} \leq ||H||_{\infty} \}$$

(Here, as usual, $\boldsymbol{B}(T)$ stands for the collection of all bounded real maps on T and $||\ .\ ||_{\infty}$ stands for the sup-norm, that is, $||g||_{\infty} = \sup \{| g(t) | : t \in T \}$ and similarly for $||H||_{\infty}$). We endow X with the pointwise partial order $\geq$,that is, $f \geq g$ iff $f(t) \geq g(t)$ for every $t \in T$. It is easy to see that $(X, \geq)$ is a complete lattice (and hence, it is, trivially, countably chain-complete). Indeed, for any non empty subset S of X and for any t in T, $\{ g(t) : g \in S \}$ is a set of real numbers bounded from above. Therefore, $f : T \rightarrow R$ is well-defined *by* $f(t) = \sup \{g(t) : g \in S\}$. Obviously, $f = \bigvee S$ and it belongs to X. As $\bigvee \phi$ equals the constant function on T that equals $||H||_{\infty}$, then, every subset of X has a $\geq$- supremum.

Conclusion: *$(X, \geq)$ is a complete lattice.*

Next, consider the operator $\Phi : X \rightarrow \boldsymbol{R}^T$ defined by

$$\Phi(g)(t) = H(t, g(h(t))), \text{ for every } t \in T.$$

(Here, as usual, $\boldsymbol{R}^T$ stands for the set of all real maps on T.) Obviously, the sup-norm of every function in the range of Φ is at most $||H||_{\infty}$, so we have $\Phi(X) \subseteq X$, that is, Φ is a self - map on X. Besides, there is a map g_* in X with $\Phi(g_*) \geq g_*$ for instance, consider the constant map $t \mapsto -||H||_{\infty}$. Finally, note that if (g_m) is a sequence in X with $\ldots\ldots \geq g_2 \geq g_1$, then (g_m) converges to $\bigvee \{g_1, g_2, \ldots\ldots\}$ pointwise and as H is increasing and continuous in its second component,

$$\begin{aligned}\sup \Phi(g_m)(t) &= \sup H(t, g_m(h(t))) \\ &= H(t, \sup g_m(h(t))) \\ &= \Phi(g)(t)\end{aligned}$$

for each $t \in T$, establishing that Φ is $\geq$- continuous.We can now invoke the Tarski-Kantorovitch Theorem to conclude that Φ has a fixed point. But this means that there is a g in X such that (2) holds, that is, there is a solution to our functional equation in X. Furthermore, again by the Tarski-Kantorovitch Theorem, we see that the collection of all such solutions forms a countably chain-complete poset under the partial order $\geq$ and possesses a $\geq$- minimum element.

Corollary on page 129 can be used to solve certain systems of functional equations in a similar way.

4.3.3(b) Application: The Contraction Mapping Theorem

Theorem : Let (X, d) be a complete metric space and f a self-map on X with

$$d(f(x), f(y)) \leq \psi(d(x, y))d(x, y) \text{ for every } x, y \in X\text{....(3)}$$

where $\psi : \boldsymbol{R}_+ \rightarrow [0, 1)$ is a continuous and decreasing function. Then, f has a unique fixed point.

Our objective here is to deduce this result from the Tarski-Kantorovitch Theorem by "embedding our metric space in a poset".

Let (X, d) be a metric space and consider the binary relation $\succeq_d$ on $X \times \boldsymbol{R}_+$ defined by

$$(x, a) \succeq_d d(y, b) \text{ if and only if } d(x, y) \leq b - a$$

Obviously, $\succeq_d$ is reflexive. It is also antisymmetric because $(x, a) \succeq_d d(y, b) \succeq_d d(x, a)$ means $d(x, y) \leq \min\{a - b, b - a\}$ which is possible only if $d(x, y) = 0$ and $a = b$. Finally, $\succeq_d$ is transitive because $(x, a) \succeq_d d(y, b) \succeq_d d(z, c)$ implies

$$\begin{aligned} d(x, z) &\leq d(x, y) + d(y, z) \\ &\leq b - a + c - b \\ &= c - a \end{aligned}$$

so that $(x, a) \succeq_d d(z, c)$. It is in this sense that we say:
Every metric space can be embeddedin a poset.

We shall proceed in after stating few lemmas.

Lemma A: *Let (X, d) be a metric space and $((x_m, a_m))$ an $\succeq_d$ -increasing sequence in $X \times \mathbf{R}_+$. Then, (x_m) is a Cauchy sequence in (X, d) and (a_m) converges.*

Lemma B : *Let (X, d) be a complete metric space and $((x_m, a_m))$ an $\succeq_d$ - increasing sequence in $X \times \mathbf{R}_+$. Then, for any $x \in X$ and $a \geq 0$,*

$$(x, a) = \bigvee \{ (x_1, a_1), (x_2, a_2), \ldots \}$$

if and only if $x_m \to x$ and $a_m \downarrow a$. In particular, $(X, \succeq_d)$ is a countably chain-complete poset.

Lemma C : *Let (X, d) be a complete metric space and f a self-map on X such that (3) holds for some continuous and bounded self-map ψ on $\mathbf{R}_+$. Then, the self-map F on $X \times \mathbf{R}_+$ defined by*

$$F(x, a) = (f(x), \ \psi(a)a),$$

is $\succeq_d$ - continuous.

Proof of the theorem : It is obvious that f can have at most one fixed point for (3) would fail for two distinct fixed points x and y of f. To prove existence of a fixed point, we pick an arbitrary element x_* of X and choose any real number $b > 0$ large enough that d

$(x_*, f(x_*)) < (1-\psi\ (b))b$. Then, we have $F(x_*, b) = (x_*, b)$. In account of *Lemmas A* and *C*, therefore, we may apply the Tarski-Kantorovitch Theorem to find an (x, a) in $X \times \boldsymbol{R}_+$ such that $(x, a) = F(x, a)$, which means $x = f(x)$ and $a = 0$ by definition of F.

It is, in fact, not difficult to provide a direct proof for the theorem (by showing that $(f^m(x)$ is Cauchy in (X, d) for any $x \in X$). Our focus here is instead on the"method" as opposed to the "result".

4.4 Tarski's Fixed Point Theorems

4.4.1 The Knaster-Tarski Theorem

As useful as the Tarski-Kantorovitch Theorem is, its order-continuity requirement is not met in a variety of applications of order-theoretic fixed point theory. Indeed, in most cases we know nothing about the self-map at hand other than that it is order-preserving. The following fundamental result shows that we can also handle such cases with ease, provided that the poset under consideration behaves suitably.

Tarski's Fixed Point Theorem : *Let $(X, \succeq)$ be a conditionally complete poset with a $\succeq$- minimum and a $\succeq$- maximum. Then, or every $\succeq$- preserving self-map f on X, $Fix(f)$ contains a $\succeq$- minimum and a $\succeq$- maximum element.*

The Knaster-Tarski Theorem: *Let $(X, \succeq)$ be a complete lattice and f a $\succeq$- preserving self-map on X. Then, $(Fix(f), \succeq)$ is a complete lattice.*

4.4.2 Application: The Schröder - Bernstein Theorem

Recall that two nonempty sets A and B are thought of as "equally crowded" if the elements of these sets can be put in one-to-one correspondence with each other. This way of looking at things allows us to compare the contents of two infinite sets and conforms fully with intuition in the case of finite sets. (To wit, two finite sets are “equally crowded” iff they contain the same number of elements, and no two sets, one finite and the other infinite, is “equally crowded”.)

A logical next step is, of course, to say that A is “less crowded” than B, if there exists an injection from A into B. After all, this means that A and a subset of B are “equally crowded”. (For instance, $\boldsymbol{N}$ is “less crowded” than $\boldsymbol{Z}$ and $\boldsymbol{R}$). We would then like to declare A and B as “equally crowded” iff A is “less crowded” than B and conversely, B is “less crowded” than A. This is true indeed, but it is hardly a trivial fact. It is instead one of the major findings of Cantorian set theory. It goes by the name.

The Schröder-Bernstein Theorem : *Let A and B be two nonempty sets such that there is an injecttion from A into B, and also, there is an injection from B to A. Then, there is a bijection from A onto B.*

It turns out that one can deduce this result from the Knaster-Tarski Theorem. We first derive the following intermediate step which makes the logic of the argument transparent.

Banach's Decomposition Theorem : *Let X and Y be two nonempty sets and take any two functions* $f : X \rightarrow Y$ *and* $g : Y \rightarrow X$*: Then, X can be written as the union of two disjoint sets* X_1 *and* X_2 *and Y can be written as the union of two disjoint sets* Y_1 *and* Y_2*, such that* $f(X_1) = Y_1$ *and* $g(Y_2) = X_2$.

Proof: We define the self-map F on 2^X by

$$F(S) = X \setminus g(\setminus f(S))$$

and observe that $B \supseteq A$ implies $F(B) \supseteq F(A)$ for any A and B in 2^X. We may, thus, apply the Knaster - Tarski Theorem (with respect to $(2^X, \supseteq)$ and F) to conclude that there exists a set $S \subseteq X$ such that $F(S) = S$. Then $g(Y \setminus f(S)) = X \setminus S$ and hence letting $X_1 = S$, $X_2 = X \setminus S$, $Y_1 = f(S)$ and $Y_2 = Y \setminus f(S)$ completes the proof.

The Schröder-Bernstein Theorem is an easy corollary of Banach's Decomposition Theorem.

Proof of the Schröder-Bernstein Theorem : Let A and B be two nonempty sets and take any two injections $f : A \to B$ and $g : B \to A$. By Banach's Decomposition Theorem, there exist a partition $\{ A_1, A_2 \}$ of A and a partition $\{B_1, B_2\}$ of B such that $F = f|_{A1}$ is a bijection from A_1 onto B_1, and $G = g|_{B2}$ is a bijection from B_2 onto A_2. But, then, $h : A \to B$, defined by

$$h(x) = \begin{cases} F(x), & \text{if } x \in A_1 \\ G^{-1}(x), & \text{if } x \in A_2 \end{cases}$$

is a bijection and we are done.

4.5 The Abian-Brown Fixed Point Theorem

4.5.1 The Abian-Brown Theorem

In many applications the domain of the equation we wish to solve is not a complete lattice, so the Knaster-Tarski Theoremdoes not apply. In addition, the Tarski-Kantorovitch Theorem often fails to remedy this situation, because this result is based on a demanding order-continuity requirement. In this section, we shall instead use the Bourbaki-Witt Theorem to obtain a fixed point theorem that will improve our standing on both of these fronts. Put simply, our objective is to replace the "finiteness" requirement in the proposition of page 126 with "chain - completeness".

Lemma : *Let $(X, \succeq)$ be a chain-complete poset with a $\succeq$- minimum and f a $\succeq$ - preserving self-map on X such that $f(x) \succeq x$ for every $x \in X$. Then, there is a $\succeq$- minimum element of $Fix(f)$.*

The Abian-Brown Theorem : *Let $(X, \succeq)$ be a chain-complete poset and f a $\succeq$- preserving self - map on X. If there exists an x_* in X with $f(x_*) \succeq x_*$, then $Fix(f) \bigcap x_*^{\uparrow}$ has a $\succeq$ - minimum and a $\succeq$- maximal element.*

Corollary : *Let* $(X, \succeq)$ *be a chain-complete poset that has a* $\succeq$*-minimum and* f *a* $\succeq$*-preserving self-map on X. Then,* $(Fix(f), \succeq)$ *is a chain-complete poset with a* $\succeq$*-minimum.*

Corollary : *Every chain-complete poset with a minimum element has the fixed point property.*

Proposotion : *Let* $(X \times \theta, \succeq)$ *be the product of a chain-complete poset* $(X, \succeq_X)$ *that has a* $\succeq_X$*-minimum and a poset* $(\theta, \succeq_\theta)$*. If* $f : X \times \theta \to X$ *is an order-preserving map, then*

$$Fix(f(\,.\,, \alpha)) \succeq_X Fix(f(\,.\,, \beta))$$

for every α *and* β *in* θ *with* $\alpha \succ_\theta \beta$.

Proposition : *Let* $(X, \succeq)$ *be a chain-complete poset and* $\mathcal{F}$ *a commutative collection of* $\succeq$*-preserving self-maps on X. If there exists an* x_* *in X with* $f(x_*) \succeq x_*$ *for each* $f \in \mathcal{F}$*, then* $Fix(\mathcal{F}) \cap x_*^{\uparrow}$ *has a* $\succeq$*-minimum and a* $\succeq$*-maximal element.*

Corollary : *Let* $(X, \succeq)$ *be a poset in which every* $\succeq$*-chain has an* $\succeq$*-infimum in X. If X has a* $\succeq$*-maximum and* f *is a* $\succeq$*-preserving self-map on X, then Fix(f) has a* $\succeq$*-maximum and a* $\succeq$*-minimal element.*

4.5.2 An Application to Fixed Set Theory

Fixed sets of self-maps, that is, those nonempty sets that are equal to their images under a given self-map appear in various branches of mathematics. For instance, in dynamical systems, the global attractors of a semi flow is defined as a maximal attracting set which is fixed under every member of the flow

In this subsection, we shall prove a basic existence theorem for

Compact fixed sets of certain forms of "continuous" self correspondences as an application of the Abian-Brown Theorem. Let us first clarify what we mean by a "fixed set" here.

Definition : *Let Γ be a self-correspondence on a nonem- pty set X. A nonempty subset S of X is called a fixed set of Γ if $S = \Gamma(S)$, that is,*

$$S = \bigcup\{ \Gamma(x) : x \in S\}$$

An easy application of the Abian-Brown Theoremshows that every self-correspondence Γ on a compact topological space X is sure to possess a largest compact fixed set, provided that Γ is a closed self-correspondence in the sense that it maps closed subsets of X to closed subsets of X, that is, $\Gamma(S)$ is closed in X for every closed subset S of X.

Theorem : *Let X be a compact topological space, and Γ a closed self-correspondence on X. Then, there exists a $\supseteq$ - maximum compact fixed set of Γ.*

Proof : Let $c(X)$ stand for the collection of all nonempty closed subsets of X, and define the map $f: c(X) \to 2^X$ by

$$F(S) = \Gamma(S)$$

As Γ is a closed correspondence, we have $f(c(X)) \subseteq c(X)$,that is, f is a self-map on $c(X)$. Moreover, obviously, $(c(X), \supseteq)$ is a poset with a $\supseteq$- maximum (namely, X), and f is $\supseteq$- preserving. It is also easy to see that every $\supseteq$- chain in $c(X)$ has an $\supseteq$- infimum in $c(X)$. Indeed, if S is a $\supseteq$- chain in $c(X)$, then S is a collection of closed subsets of X with the .nite intersection property. Therefore, as X is compact, we have $\bigcap S \neq \phi$ hence $\bigcap S$ is the $\supseteq$- infimum of S in $c(X)$.

We now apply the last corollary of the previous page to the poset $(c(X), \supseteq)$ and the self-map f to find that Fix(f) has a $\supseteq$- maximum element. But, as every closed subset of X is compact (because X is compact), Fix(f) is exactly the collection of all compact fixed sets of Γ and we are done.

A natural question is if we could also prove that the collection of all compact fixed sets of Γ constitute a complete lattice (with respect to $\supseteq$) under the conditions of the above theorem. We cannot invoke the Knaster-Tarski Fixed Point Theorem in the argument we gave above to do this, because $(c(X), \supseteq)$ is, in general, not even a lattice.

Thanks For Reading

I share words of thanks to the esteemed readers of this book. I believe this simple book will benefit the reader a lot. I welcome suggestion from the readers so as to bring out this book with a better taste in the future edition.

Ph. Suranjoy Singh

Bibliography

A. Wright, *Sperner's Lemma and Brower's Fixed Point Theorem*, www..uchicago.edu/~amwright/BFPT.

Brower, Nash, www.math.uchicago.edu/~shmuel/AAT-readings/ Econ% 20seg- ment.

C. Melcher, R. Aachen, *Nonlinear Analysis : Fixed Point Theory*, www.math1.rwth-aachen.de/files/NA/NA1_11_Lec_0

C. Rousseau, *Banach Fixed Point Theorem and applications*, September 12, 2010, University of Montreal.

C.U. Santa, A.R. Prat, *The Lefschetz-Hopf Fixed Point Theorem,* www.staff.science. uva.nl/ ~heinloth/StacksSeminar/AndreaCamilaLefschetz.

Chapter 5, *Order-Theoretic Fixed Point Theory,* www.files.nyu.edu/eo1/public/ Book-PDF.

F.F. Bonsall, *Lectures on Some Fixed Point Theorems on Functional Analysis*, TIFR Bombay, 1962.

J. Jachymski, *Order-Theoretic Aspects of Metric Fixed Point Theory*, Handbook of Metric Fixed Point Theory, 2001, pp 613-641 (Springer link).

K. Conrad, *The Contraction Mapping Theorem*, www.math.uconn.edu/~kconrad/ blurbs/ analysis/contraction.

M.A. Khamsi, *Introduction to Metric Fixed Point Theory*, National Workshop on Nonlinear Functional Alalysis and its Applications, Shahid Beheshti University, January 20-24, 2002.

O. Shalit, *Advanced Analysis, Notes 12: Banach spaces (application: existence of Haar measure)*, Noncommutative Analysis, www.noncommutativeanaly-sis.wordpress. Com.

P. Hitzler, A.K. Seda, *A "Converse" of the Banach Contraction Mapping Theorem*, www.citeseerx.ist.psu.edu/viewdoc/summary?doi=10.1.1.29.736.

P. Kumlin, Mathematics, Chalmers & GU, *A Note On Fixed Point Theorey*, TMA 401/MAN 670, Functioanl Analysis 2003/2004.

R.P. Agarwal, M. Meehan and D. O'regan, *Fixed Point Theory and Applications*, Cambridge University Press.

T. Stuckless, *Brower's Fixed Point Theorem : Methods of Proof and Generalization*, www.oldweb.cecm.sfu.ca/Preprints03/2003-207.

V. Pata, *Fixed Point Theorems and Applications*, Dipartimento di Matematica "F. Brioschi" Politecnico di Milano, www.mate.polimi.it/viste/pagina_personale/ pp/121/fp.

X. Chen, B. Kiskovski and M. Kambadur, *Lecture 4, Nash's Mixed Equilibrium and Brouwer's Fixed Point*, COMS E6998: Algorithmic Game Theory and Eco -nomics Feb 10, 2011.